RECOMMENDER SYSTEMS

RECOMMENDER SYSTEMS

Algorithms and Applications

Edited by
P. Pavan Kumar
S. Vairachilai
Sirisha Potluri
Sachi Nandan Mohanty

CRC Press
Taylor & Francis Group
Boca Raton London New York

CRC Press is an imprint of the
Taylor & Francis Group, an **informa** business

First Edition published 2021
by CRC Press
6000 Broken Sound Parkway NW, Suite 300, Boca Raton, FL 33487-2742

and by CRC Press
2 Park Square, Milton Park, Abingdon, Oxon, OX14 4RN

ISBN: 978-0-367-63185-7 (hbk)
ISBN: 978-0-367-63187-1 (pbk)
ISBN: 978-0-367-63188-8 (ebk)

Typeset in Caslon
by SPi Global, India

Contents

Preface

Recommendations in specific domains and milieus can be regarded as vital sources that have a significant impact on the recommendation goals. Different types of data, such as content data, historical data and user data based on behavior and views, are all used to fine-tune the training models in recommender systems. Similarly, in an industrial perspective, individual item or product recommendation is enhanced to promote sales. Recommender systems are targeted at predicting recommended products in order to provide personalized results in specific domains and fields. The context of a recommendation system can be viewed as ensuring a decisive classification that affects the recommendation goals. Recommender systems search through large and expansive amounts of data, such as temporal, spatial, and social, by using potential prediction techniques to produce the most accurate recommendations. From an industry point of view, both service providers and product users benefit from improved decision-making processes and accurate recommendation techniques. There are a wide range of robust recommender systems, including, among others, the Trust-Based Recommender System, the Real-Time Personalized Recommender System, and High Accuracy, and Diversity-Balanced Recommender Systems. These have been used in an increasingly diverse range of fields, including Information Technology, Tourism,

Education, E-Commerce, Agriculture, Health Care, Manufacturing, Sales, Retail, and many more.

Various efficient and robust product recommender systems using machine learning algorithms have proved effective in filtering and exploring unseen data by users for better prediction and extrapolation of their decisions. These are providing a wider range of solutions for various problems, particularly those involving imbalanced dataset problems, cold-start problems, and long tail problems. Community detection algorithms on many datasets, such as temporally evolving datasets, are also being seen as the most capable algorithms in various domains. Ontology-based knowledge representation and reasoning techniques are the best for representing sanctions and supporting better reasoning. Collaborative filtering-based recommender systems make extensive use of latent factor models to solve machine learning problems using benchmark datasets. Collaborative and content-based approaches to social networking content are also generating strong, resilient, and meaningful suggestions about new products and services. In another example, using the Tropos Goal-Risk Model, risk assessment and analysis for a particular software product is developed in order to satisfy all perspectives of recommendations. Contemporary and imminent research in the area of recommender systems using various algorithms and applications delivers opportunities to a large number of areas and domains.

Acknowledgements

The editors would like to congratulate and thank all of the authors for their contributed chapters: Utkarsh Pravind, Palak Porwal, Abhaya Kumar Sahoo, Chittaranjan Pradhan, B.S.A.S. Rajita, Mrinalini Shukla, Deepa Kumari, Subhrakanta Panda, Vidya Kamma, Dr. D. Teja Santosh, Dr. Sridevi Gutta, Aleena Mishra, Mahendra Kumar Gourisaria, Palak Gupta, Sudhansu Shekhar Patra, Lalbihari Barik, R. S. M. Lakshmi Patibandla, V. Lakshman Narayana, Arepalli Peda Gopi, B. Tarakeswara Rao, Dr. G. Ramesh, Dr. P. Dileep Kumar Reddy, Dr. J. Somasekar, R. Bhuvanya and Ms. Pallavi Mishra.

We would also like to thank the subject matter experts who found the time to review the chapters and deliver those in time. Finally, a ton of thanks to all the team members of CRC Press/Taylor & Francis Group for their dedicated support and help in publishing this edited book.

Editors

Dr. P. Pavan Kumar received his PhD from JNTU Anantapur, India, in 2019, and his M.Tech. from NIT, Tiruchirappalli with MHRD scholarship from Govt of India. Currently, he is working as an Assistant Professor in the Department of Computer Science and Engineering at ICFAI Foundation for Higher Education (IFHE), Hyderabad. His research interests include real-time systems, multi-core systems, high-performance systems and computer vision. He is the recipient of 1 Best Paper award during his PhD from International Conference at PEC, Pondicherry, in 2017. In addition to research, he has guided two M.Tech. students and 25 B. Tech. students. He has published around 20 scholarly peer reviewed research articles in reputed international journals as well as presented his work in various international conferences and two Indian patents. He is a life member of ACM-CSTA and IAENG.

Dr. S. Vairachilai was awarded her PhD in Information Technology from Anna University, India, in 2018. She received her M.Tech, from Kalasalingam University, India, and MCA from Madurai Kamaraj University, India. She is currently working as an Associate Professor in the Department of CSE at CVR College of Engineering, Hyderabad, Telangana, since 2021 (March). Prior to this, she served in teaching roles from ICFAI Foundation for Higher Education (IFHE), Kalasalingam University, and N.P.R College of Engineering and Technology, Tamilnadu, India. Her research interests include Machine Learning, Data Science, Recommender Systems and Social Network Analysis. She has authored more than 30 scholarly peer reviewed research articles, which are published in presumed national and international Journals and Conferences. She has four patents and occupied as PhD supervisor for Annamalai University to guide two research scholars in the machine learning domain. She is a reviewer for Elsevier (*Computer Communications & Microelectronics Journal*) and Life Member of "The Indian Society for Technical Education" (ISTE).

Ms. Sirisha Potluri is working as Assistant Professor in the Department of Computer Science & Engineering at ICFAI Foundation for Higher Education, Hyderabad. She is pursuing her PhD in the area of cloud computing. Her research areas include distributed computing, cloud computing, fog computing, recommender systems and IoT. She has 7+ years of teaching experience. Her teaching subjects include Computer Programming using C, Data Structures, Core JAVA, Advanced JAVA, OOP through C++, Distributed Operating System, Human Computer Interaction, C# and .NET Programming, Computer Graphics, Web Technology, UNIX Programming, Distributed and Cloud Computing, Python Programming and Software Engineering. She has published 15 scholarly peer reviewed research articles in reputed international journals. She has attended 12 international conferences organized by various professional bodies like IEEE and Springer to present her novel research papers. She has a patent in cloud computing domain. She is a Member of IEEE Computer Society. Sirisha is a reviewer for various journals of Inderscience Publishers and has reviewed many manuscripts.

 **Dr. Sachi Nandan Mohanty** is an Associate Professor in the Department of Computer Engineering at the College of Engineering Pune, India. He received his PostDoc from IIT Kanpur in 2019 and PhD from IIT Kharagpur, India, in 2015, with MHRD scholarship from Govt of India. He has edited 14 books in association with Springer and Wiley. His research areas include Data Mining, Big Data Analysis, Cognitive Science, Fuzzy Decision Making, Brain-Computer Interface, Cognition, and Computational Intelligence. Prof. Mohanty has received 3 Best Paper Awards during his PhD at IIT Kharagpur from International Conference at Beijing, China, and at the International Conference on Soft Computing Applications organized by IIT Rookee in 2013. He was awarded Best thesis award (first prize) by the Computer Society of India in 2015. He has published 30 International Journals of international repute and has been elected as FELLOW of Institute of Engineers and Senior member of IEEE Computer Society Hyderabad chapter. He is also the reviewer of *Robotics and Autonomous Systems* (Elsevier), *Computational and Structural Biotechnology Journal* (Elsevier), *Artificial Intelligence Review* (Springer), and *Spatial Information Research* (Springer).

Contributors

S. Anusha
KSN Government Degree College
  for Women
Ananthapuramu, Andhra Pradesh,
  India

Lalbihari Barik
Department of Information
  Systems
Faculty of Computing and
  Information Technology in
  Rabigh
King Abdul Aziz University
Jeddah, Kingdom of Saudi Arabia

R. Bhuvanya
Vel Tech Rangarajan Dr. Sagunthala
  R&D Institute of Science and
  Technology
Chennai, Tamil Nadu, India

Arepalli Peda Gopi
Department of Computer Science
  and Engineering
Vignan's Nirula Institute of
  Technology & Science for
  Women
Guntur, Andhra Pradesh, India

Mahendra Kumar Gourisaria
School of Computer Engineering
KIIT Deemed to be University
Bhubaneswar, Odisha, India

Palak Gupta
School of Computer Engineering
KIIT Deemed to be University
Bhubaneswar, Odisha, India

Sridevi Gutta
Department of CSE Gudlavalleru
Engineering College Gudlavalleru
Vijayawada, Andhra Pradesh, India

Vidya Kamma
Department of CSE
KL University
Guntur, Andhra Pradesh, Inida

M. Kavitha
Vel Tech Rangarajan Dr. Sagunthala
    R&D Institute of Science and
    Technology
Chennai, Tamil Nadu, India

Deepa Kumari
Department of CSIS
BITS-Pilani
Hyderabad, Telangana, India

Aleena Mishra
School of Computer Engineering
KIIT Deemed to be University
Bhubaneswar, Odisha, India

Pallavi Mishra
Department of Computer Science &
    Engineering, Faculty Of Science
    & Technology (FST)
ICFAI Foundation For Higher
    Education, Hyderabad
Telangana, India

V. Lakshman Narayana
Department of Information and
    Technology
Vignan's Nirula Institute of
    Technology & Science for
    Women
Guntur, Andhra Pradesh, India

Subhrakanta Panda
Department of CSIS
BITS-Pilani
Hyderabad, Telangana, India

R. S. M. Lakshmi Patibandla
Department of Information and
    Technology
Vignan's Foundation for Science,
    Technology and Research
Guntur, Andhra Pradesh, India

Sudhansu Shekhar Patra
School of Computer Applications
KIIT Deemed to be University
Bhubaneswar, Odisha, India

Palak Porwal
School of Computer Engineering
KIIT Deemed to be University
Bhubaneswar, Odisha, India

Chittaranjan Pradhan
School of Computer Engineering
KIIT Deemed to be University
Bhubaneswar, Odisha, India

Utkarsh Pravind
School of Computer Engineering
KIIT Deemed to be University
Bhubaneswar, Odisha, India

B.S.A.S. Rajita
Department of CSIS
BITS-Pilani
Hyderabad, Telangana, India

B. Tarakeswara Rao
Department of Computer Science
    and Engineering
Kallam Haranadhareddy Institute of
    Technology
Guntur, Andhra Pradesh, India

G. Ramesh
GRIET
Hyderabad, Telangana, India

P. Dileep Kumar Reddy
S.V. College of Engineering
Tirupati, Andhra Pradesh, India

Abhaya Kumar Sahoo
School of Computer Engineering
KIIT Deemed to be University
Bhubaneswar, Odisha, India

D. Teja Santosh
Department of IT
Sreenidhi Institute of Science and
    Technology
Ghatkesar, Telangana, India

Mrinalini Shukla
Department of CSIS
BITS-Pilani
Hyderabad, Telangana, India

J. Somasekar
Gopalan College of Engineering and
    Management
Bangalore, Karnataka, India

1

COLLABORATIVE FILTERING-BASED ROBUST RECOMMENDER SYSTEM USING MACHINE LEARNING ALGORITHMS

UTKARSH PRAVIND,
PALAK PORWAL,
ABHAYA KUMAR SAHOO
AND CHITTARANJAN PRADHAN

KIIT Deemed to be University, Bhubaneswar, Odisha, India

Contents

1.1 Introduction

In this chapter, we will introduce a number of different machine learning algorithms, such as K-Nearest Neighbors (KNN), SVM and Random Forest. All of these algorithms use collaborative filtering technology. It has two paths – a narrow one and another, more general one. In the narrower one, automatic predictions about the interests and the preferences of the user are collected by collaborative filtering pieces of information from several other users; contrastingly, in more general filters for relevant information by collaboration among multiple users, data sources, etc. [1–3]. There are various types of collaborative filtering based on the nature of interaction i.e. Explicit feedback and Implicit feedback. Another classification is drawn between memory-based and model-based filters. Collaborative filtering has many challenges, such as data sparsity, scalability, synonyms, gray sheep, diversity, and long tails. These Filtering techniques are used along with the Machine Learning model in this chapter we will learn about three models: KNN, SVM, and Random Forest [1, 4, 5].

K-Nearest Neighbors (KNN) is one of the most basic algorithms and is used in both regression and classification problems. KNN uses data and classifies new data points based on measures of similarity. KNN works by calculating the distance between all examples in the data and the query, select the K-nearest neighbours to the query and vote for the most frequent label. SVM is a supervised machine learning model with related learning algorithms that analyze data for classification and regression analysis. In SVM, the main goal is to find the hyperplane in the N-dimensional space with the largest margin. SVM has better outliers than KNN. In cases where the training data are much larger than the number of features (m>>n), then KNN is better than SVM. However, SVM outperforms KNN when there are large features and there are less training data. The Random Forest consists of many decision trees, which it makes by using features like bagging and randomness. Random Forest is the supervised learning algorithm that is used in both classifications as well as regression. It is changeable, and used easy-to-use machine learning algorithms that produce without NY hyper-parameter tuning. All these algorithms can be used to build a Recommendation system [6–8].

A recommender system offers useful suggestions. Such systems are used in various areas; for example, when we want to play a video or program on platforms such as YouTube or Netflix, or a song on Spotify. Similar systems are employed on online retail channels such as Amazon or Flipkart or on social networking platforms such as Facebook, Twitter, and Instagram [9].

In our study when we build a recommendation system using the above machine learning algorithm, we found that, compared with the accuracy of SVM and Random Forest, KNN is the most accurate. Throughout this chapter, we will learn about different filtering techniques of the Recommender System, which are used to filter out the relevant information about the user's preference. In what follows we will also consider their relative advantages and disadvantages. Next, we will study some of the different methods of recommendation system used in machine learning, namely KNN, SVM, Random Forest, and decision trees [10, 11]. The different methodologies each have their own pros and cons. KNN uses different metrics to evaluate data, while Random Forest is a collection of decision trees. We will learn about how to calculate the importance of features and its implementation in Spark and scikit-learn with several other terminologies like Information Gain, Node Impurity. There are some limitations to these models such as so-called 'cold start' problems, which occur because of a lack of information about the customers or items, meaning that making accurate predictions is sometimes difficult [12–14].

And, lastly, we will go discuss different domains in which the recommendation system can be used as a powerful tool for shaping a better future.

1.2 The Background of the Recommender System

The recommendation system is based on personalized information filtering that seeks to predict the "preference" or "rating" of the users mainly by building the model based on a comparison between the characteristics of an item and the users' profile (content-based filtering) and the user's domain (collaborative filtering). Different algorithms are used to implement the recommendation system, factors on which the selection criteria may depend are training set example quantity, featured space's dimensions [7, 15, 16] (Figures 1.1 and 1.2).

1.3 Types of Filtering

1.3.1 Content-Based Technique

This is also known as cognitive filtering, and is a system which suggests items based on the contrast between the content of an item and the users. It represents the items in the form of a set of descriptors [17–19].

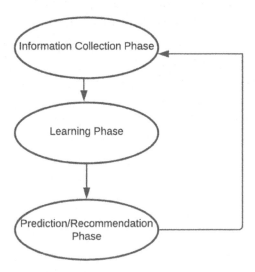

Figure 1.1   Different Phases of Recommender System

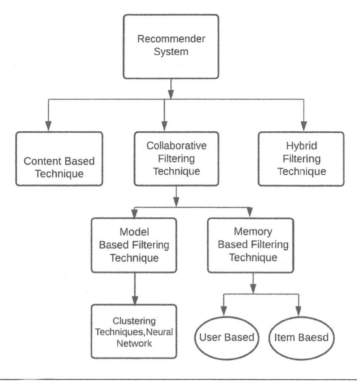

Figure 1.2 Different Approaches of Recommender System

1.3.2 Collaborative-Based Technique

Collaborative filtering, also known as social filtering, filters information by drawing on the recommendations of different people. The key idea behind this technique is that similar people share the same interests and therefore similar items will be liked by the user. As time progresses, the system will be able to give an increasingly accurate and relevant result [6, 20, 21].

1.3.3 Hybrid-Based Technique

The hybrid-based approach combines collaborative filtering, content-based filtering, and other approaches. Drawing on the advantages of all of the techniques used in it in order to give an accurate result. But it also faces newer challenges that evolve such as variation among the users, change in user taste [22, 23].

1.3.4 *Pros of Collaborative Filtering*

Independent on User: collaborative filtering needs the help of other users' recommendations to find similar preferences between them and then give the recommendation for the item. Instead, the content-based technique makes recommendations based on sthe comparison between the items of the user profile for a recommendation [24, 25].

Transparency: Some substantial numbers of users have the same likes and dislikes based on which it gives suggestion to the user but the content-based recommendation technique is user-specific.

No cold start: Even when the information of a certain product is unavailable, in collaborative filtering, it is possible to recommend new products before obtaining any customer ratings [1].

1.3.5 *Cons of Collaborative Filtering*

Limited content analysis: If the information does not carry much information to distinguish the products sharply, then it is not possible to use the recommendation engine.

Over-specialization: Content-based method provides limited modernity since it has to match features of the user's account and items. Perfect content-based filtering may indicate nothing "surprising."

New User: When the new users arrive due to lack of knowledge about the user it could not suggest any article [25] (Figure 1.3).

1.4 Different methods for Recommendation systems

1.4.1 *K-Nearest Neighbors (KNN)*

KNN is a method that does not use any parameter and is used in the implementation classification as well as regression problems. The input consists of the k-nearest training items in the feature space. KNN is based on lazy learning method in which the dataset is continuously updated with new entries.

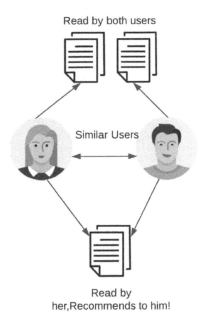

Read by both users

Similar Users

Read by
her,Recommends to him!

Figure 1.3   Collaborative Filtering

1.4.2 Support Vector Machine (SVM)

SVM is a supervised learning model which checks the data used in the classification and regression process; it is widely used in the classification process. The objective of SVM is to divide the datasets in order to find a maximum marginal hyperplane (MMH).

1.4.3 Random Decision Trees/Random Forest

Random Forest is a group learning technique for both classification as well as regression and other processes that utilize by constructing decision trees at training time and the result is the combination of the outputs of the individual trees. They are a powerful model as they limit the overfitting problem to their training set.

1.4.4 Decision Tree

Decision trees are powerful machine learning algorithms which provide the basic building blocks for other algorithms. They are used for classification as well as regression. The objective of the decision tree is

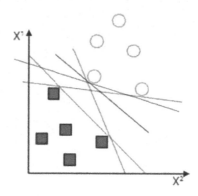

Figure 1.4　SVM (Different Hyper-Plane between 2 Types of Data)

to construct a model which predicts the value of a target variable by learning decision rules deduced from the dataset.

In statistics, the logistic model is used to find the probability of some class, events, or existing circumstances, such as success/failure, or well/sick. This can also be enlarged to model such as to determine whether an image consists of a zebra, tiger, dog, etc. If an image found of an object which we are searching then the probability is assigned between the value zero and 1. Logistic regression is a statistical model because it makes use of a logistic form to model a binary dependent variable, even though many extra complicated extensions exist.

1.5 Support Vector Machine

SVM (or Support Vector Machine) is a linear model that is used for both classification and regression models. It is capable of solving both linear and non-linear problems and work as an efficient algorithm for many practical problems. The principal idea behind SVM is simple, i.e. it creates a line or a hyperplane which segregates the data into classes [13] (Figure 1.4).

In the SVM algorithm, we basically want to increase the margin between the data points and the hyperplane. To increase the margin, we use Hinge loss function [14].

Hinge loss function:

$$c(x, y, f(x)) = \begin{cases} 0, \text{ if } y * f(x) \geq 1 \\ 1 - y * f(x), \text{ else} \end{cases}$$

$$c\left(x, y, f\left(x\right)\right) = \left(1 - y * f\left(x\right)\right) +$$

If the predicted value sign becomes the same as the actual value sign, then the cost function becomes 0.Otherwise the loss value can be calculated easily. We also add parameter to the cost function, which is the regularization parameter. The main purpose of using regularization function is to margin the maximization and the loss [14].

$$min_w \lambda \parallel w \parallel^2 + \Sigma\left(1 - y_i\left(x_i, w\right)\right) +$$

On taking the partial derivatives with weight, it helps to identify the gradients. By using gradient, we can update weight.

$$\frac{\delta}{\delta w_k} \lambda \parallel w \parallel^2 = 2\lambda w_k$$

$$\frac{\delta}{\delta w_k}\left(1 - y_i(x_i, w)\right) + = \begin{cases} 0, \textit{if } y_i(x_i, w) \geq 1 \\ -y_i x_{ik}, \textit{else} \end{cases}$$

The model rightly predicts the value, gradient of the regularization parameter can be changed.

$$\omega = \omega - \alpha \cdot \left(2\lambda\omega\right)$$

Whenever our model makes a wrong prediction of the class of our data point; we add on the loss along with the regularization parameter update our gradient.

1.6 KNN

Both classifications, as well as a regression problem, can be implemented by using the KNN algorithm. Due to its simple and easy nature, it is mostly used for implementing classification problems in the industries. In the case of non-parametric techniques, this means less knowledge of the data or we can say that it does not exist. KNN is also called a lazy algorithm either because there is no clear training phase or because the training phase requires very little time.

The entire training set holds on at the time of learning and the majority of its closest neighbors in the training set is allocated a

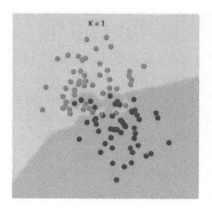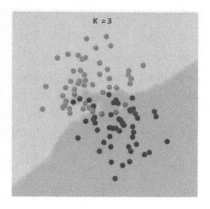

Figure 1.5 Identification of an Unknown Sample Using: (left) Only One Identified Sample; (right) More than One Identified Sample

query class. When K=1, KNN is in its simplest form. Figure 1.5(left), an unidentified sample, is classified by using only one known sample. Figure 1.5(right) more than one known sample is used. In Figure 1.5(right), the value of K is set to 3, in order to considered the closest four samples. Three quarters are based on the same category, and only one belongs to a different category.

1.6.1 Algorithm

1. First load the test data and training data.
2. Choose K value
3. For specific points in the test data:
   - Calculate the Euclidean distance of all training points
   - Euclidean distance is stored in a list, and then we sort the list
   - Select the first k points
   - According to most categories, assign a category to each test point
4. Termination

The size of K (parameter) affects the selection's sensitivity of neighborhood with K size because the radius of the local region is calculated by the distance of the Kth-nearest neighbor.

KNN performance is primarily determined by the value of a parameter K, along with the distance metric. Selection of the neighborhood size K basically affects its estimate. The distance of the Kth-nearest neighbor to the query is responsible for the radius of the local region. Many distance functions are available for KNN Euclidean distance, Squared Euclidean distance, Minkowski distance, Manhattan distance.

$$d(p,q) = (\sum_{i=1}^{n} |p_i - q_i|^c)^{1/c}$$

For $c = 1$ and $c = 2$, the Murkowski's metric becomes equal to the Manhattan and Euclidean metrics, respectively.

The KNN classifier is suitable for data streams. When a new sample appears, the K neighbors nearest is calculated from the training space based on some metrics. The number of neighbors (K) in KNN is chosen at the time of building a model [4]. The controlling variable for the prediction model is K. First inspect the data, then find the optimal value of K. In general, a large K value is more preferred as it reduces the overall noise. The optimal K for most datasets has been between 3 and 10 because it produces much better results than K=1.

1.6.2 Curse of Dimensionality

Although KNN performs well when the input variable is small, it tends to struggle when the input variable is large. In high dimensions, the similar points have very large distances among them. All points will be far away from each other.

The testing is slower and costlier in terms of time and memory. A large amount of memory is required to store the entire training set. It requires the scaling of data because Euclidean distance is taken into account between two data points in order to find nearest neighbors. Euclidean distances are sensitive to magnitudes. KNN is also not appropriate for large dimensional data.

1.7 Random Forest

Random Forest is a kind of ensemble algorithm that combines the prediction of several trees. The decision tree is an integral part of the Random Forest model.

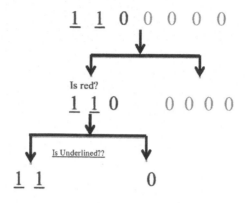

$$\underline{1} \quad \underline{1} \quad 0 \quad 0 \quad 0 \quad 0 \quad 0$$

Is red?

$$\underline{1} \quad \underline{1} \quad 0 \qquad 0 \ 0 \ 0 \ 0$$

Is Underlined??

$$\underline{1} \quad \underline{1} \qquad\qquad 0$$

Figure 1.6   Working of a Random Forest

The decision tree is one of the most approved classification techniques and is based on the 'divide and conquer' technique. A decision tree consists of a root node, branches, and leaf nodes. Inner node denotes a test on a feature, each branch denotes the result of a test carried out, and each leaf node holds a class label. The topmost level node in the decision tree is called the root/basis node (Figure 1.6).

Let us imagine a simple example. Suppose the dataset contains the following numbers, as shown at the top of Figure 1.6. There are two 1's and five 0's (1's and 0's is classes) and there is a desire to split them on the basis of these attributes. The attributes are represented as binary numbers i.e. 1 and 0. We have to find whether the given attributes are underlined or not. We identify the number: whether it is 1 or 0. If it is 1, then we check whether is is underlined or not.

1.7.1 Random Forest Classifier

Random Forests are constructed by combining the predictions of several trees, with each tree being trained individually.

The Random Forest can perform both classification and regression. Classification problems are solved using the Random Forest. Random Forest uses all classifications and selects the most suitable prediction as the result. The input of each tree is in a form of sampled data rather than the original dataset (Figure 1.7).

In the above example, the tree recommends us whether to play the game or not, based upon the weather conditions. There are three

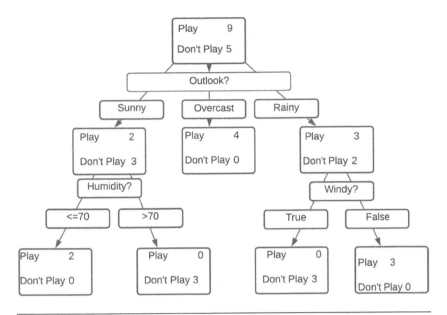

Figure 1.7 How Random Classifier works

weather (outlook) conditions, i.e. sunny, overcast or humid. If the humidity is less than or equal to 70, then it is feasible to PLAY. But if the humidity is greater than 70 then don't play the game. In the other scenario, if the weather is rainy and windy then don't play. And if the outlook is overcast than also don't play ball.

1.7.2 Mathematical Expression

1.7.2.1 Node Impurity/Impurity Criterion   scikit-learn and Spark both provide information in their documentation about the formulas that should be used for the impurity criterion. However, they all use Gini impurities by default, but they can also choose to provide entropy. scikit-learn and Spark both use Mean Square Error, for regression (Figure 1.8).

1.7.2.2 Information Gain   Information gain is another term that we need to keep in mind. After converting the dataset, this is the entropy reduction of the dataset:

$$IG\left(A,B\right) = E\left(A\right) - E\left(A,B\right)$$

Impurity	Task	Formula	Description
Gini impurity	Classification	$\sum_{i=1}^{C} f_i(1-f_i)$	f_i is the frequency of label i at a node and C is the number of unique labels.
Entropy	Classification	$\sum_{i=1}^{C} -f_i\log(f_i)$	f_i is the frequency of label i at a node and C is the number of unique labels.
Variance / Mean Square Error (MSE)	Regression	$\frac{1}{N}\sum_{i=1}^{N}(y_i-\mu)^2$	y_i is label for an instance, N is the number of instances and μ is the mean given by $\frac{1}{N}\sum_{i=1}^{N}y_i$
Variance / Mean Absolute Error (MAE) (Scikit-learn only)	Regression	$\frac{1}{N}\sum_{i=1}^{N}|y_i-\mu|$	y_i is label for an instance, N is the number of instances and μ is the mean given by $\frac{1}{N}\sum_{i=1}^{N}y_i$

Figure 1.8 Impurity Formula used by scikit learn and spark

where,
   A: targeted variable
   B: splitting feature
   E: entropy

(i) **Random Forests**
   During the training period, Random Forests (RF)build many decision trees. Predictions from each tree are combined to make a final prediction.

1.7.2.3 Implementation in scikit-learn For each decision tree, scikit-learn uses Gini importance or Mean Decrease in Impurity (MDI) to calculate the importance of a node, assuming only two child nodes (binary tree):

$$ni_j = w_j C_j - w_{l(j)}C_{l(j)} - w_{r(j)}C_{r(j)}$$

- ni_j: node j importance
- w_j : weighted samples that are reaching node j
- C_j: node j's impurity value
- $l(j)$: it is the child node of node j, from left
- $r(j)$: it is the child node of node j, from right

Feature importance is the reduction of node impurity, and the probability of the impurity reaching the node is obtained after weighting. The node probability can be calculated by dividing the number of samples arriving at the node by the total number of samples. The higher the value, the more important the function.

We can calculate importance of feature on a decision tree:

$$fi_i = \frac{\sum_{j:\text{node } j \text{ splits on features } i} ni_j}{\sum_{k \in \text{all nodes}} ni_k}$$

- fi_i= feature i's importance

Normalization of this is done by dividing the sum of all feature importance values, to a value between zero (0) and one (1):

$$normfi_i = \frac{fi_i}{\sum_{j \in \text{all features}} fi_j}$$

From the formula, it is seen that, the feature importance of Random Forest depends up on the average of all trees i.e. calculated as:

$$Rfi_i = \frac{\sum_{j \in \text{all tress}} norm\, fi_j}{t}$$

Rfi_i = It gives the feature importance (i), from all the trees

t: total trees

$norm\, fi_j$ = i's normalized features importance of node j

1.7.2.4 Implementation in Spark For each decision tree, features importance is calculated as:

$$fi_i = \sum_{j:\text{nodes } j \text{ splis on features } i} s_j C_j$$

- fi_i: feature *i*'s importance
  s_j: total samples reaching node j
  C_j : node j's impurities value

For calculating the final feature importance firstly, the importance of each tree calculates and then it is normalized to the tree:

$$normfi_i = \frac{fi_i}{\sum_{j \in all\ features} fi_j}$$

$normfi_i$: i's normalized features importance of node j

- fi_i = feature i's importance

Then the feature importance values from each tree are summed normalized:

$$Rfi_i = \frac{\sum_j normfi_j}{\sum_{j \in all\ features, k \in all\ tress} normfi_{jk}}$$

Rfi_i: This gives the feature importance (i), from all the trees
$normfi_{jk}$: i's normalized features importance of node j

1.8 Comparison of Algorithms

KNN	SVM	RANDOM FOREST
It is robust to noisy training data and is effective in the case of a large number of training examples.	The main reason to use an SVM instead is because the problem might not be linearly separable. In that case, we have to use SVM with a nonlinear kernel (e.g. Radial Basis Function).	Random Forest is nothing more than a collection of decision trees combined. They can handle categorical variables very well.
For this algorithm, we need to determine value the parameter K (Nearest neighbour) and the type of distance to use. The calculation time is also very much needed to calculate the distance of each All query examples Training samples.	Another reason to use this algorithm if we are in High-dimensional space. For example, SVM has Observed to work efficiently for Text classification. However, SVM requires a lot of time for training. and so, not recommended When we have a big Training example.	This algorithm can handle high-dimensional spaces as well as the large number of training examples.

1.9 Limitation and Challenges

1.9.1 Data Sparsity

Today, commercial recommendation systems use a large number of datasets. For this reason, the user item matrix used for collaborative filtering becomes large and complicated, which reduces the performance of the recommendation system. This problem leads to cause cold start problem which is also common in the recommendation system.

1.9.2 Scalability

Due to an increase in the number of users and items collaborative filters go through a serious scalability problem. If there is a large number of customers and a large number of items this makes the collaborative matrix complex which leads to a degradation of the performance.

1.9.3 Cold Start Problem

This is a familiar problem in the recommendation system. It happens due to a lack of information about the customers or items. Due to which the system couldn't predict correctly because no proper history of the user is available and in case of the item no ratings are available which causes inaccurate predictions. To avoid this problem, we use content-based filtering. Content-based methods provide accurate predictions in case of the new item as they don't depend on any earlier rating information of other users to recommend the item.

1.10 Future Scope

The recommender system has become a booming area for research in recent years. Although recommender systems have seen unequal improvements in primitive content-based as well as collaborative filtering methods, a lot of research work is in progress to improve the percentage accuracy of the output and improvements in all its fields. Some of the future areas in which the Recommender system has vast scope and efforts are mentioned below.

1.10.1 Medical

For building a medical system, the availability of input data is sparse as medical history is often treated as personal, confidential information, so sometimes it becomes difficult to build a reliable recommendation system. But medical recommendation engines can be used to fill the gap and support the decision-making process and to provide better healthcare.

1.10.2 Domain Adaption

The single-domain recommendation system generally focuses on one particular domain and ignores the interest of the user in other domains, leading to cause sparsity and a cold start problem. Several works are ongoing which indicate the efficacy of deep learning in catching the generalizations and producing a better recommendation system on cross-domain platforms.

1.10.3 Data Sparseness & Cold Start

Data sparseness is a major issue, making the recommendation algorithms useless. One of the other most frequently occurring problems is the cold start problem, that is becoming an impediment to the establishment of accurate recommendation systems. Many kinds of research are currently being undertaken in attempts to eliminate it.

1.11 Conclusion

Recommendation systems are a subcategory of information-filtering systems that try to predict the "level" or "preference" that users will give to the product. They are mainly used for commercial applications. In this chapter, we used three different machine learning algorithms (KNN, SVM, and Random Forest), which have been implemented, evaluated, and compared. Different datasets, including both balanced and unbalanced, have different training sample sizes. All of the classification results are very high, and between 90% and 95%. The SVM classifier produces the highest value on average and has the lowest sensitivity to the training sample size, followed by the RF and KNN classifiers. However, the difference between the various training sample sizes of the SVM classifier is not readily apparent. For all three classifiers, when the training samples are large enough, regardless of whether the dataset is unbalanced or balanced, the overall level of accuracy is similar and high (over 93.85%).

1.12 Contribution

There has been a major shift of internet appliances from the devices to the users. The major goal of recommendations is provided by the recommender system to help in better prediction, understanding the likes/dislikes of the user. At the time of writing this chapter, we faced many difficulties but continued hard work and dedication allowed us to complete the work in due time. Before directly writing this contribution, we learned the basics of all models (KNN, SVM and Random Forest) and how they work. We made a predictive model and calculated the accuracy and performance of each model that will be used. Before completing this article, we carefully read its template to understand how to write it perfectly.

References

[1] S. B. Imandoust, and M. Bolandraftar. "Application of k-nearest neighbor (knn) approach for predicting economic events: Theoretical background". *International Journal of Engineering Research and Applications*, 3(5), 605–610 (2013).

[2] G. Ruffo, and R. Schifanella. "*Evaluating peer-to-peer recommender systems that exploit spontaneous affinities.*" In *Proceedings of the 2007 ACM Symposium on Applied Computing*, ACM, pp.1574–1578 (2007).

[3] I. Avazpour, T. Pitakrat, L. Grunske, and J. Grundy. "*Dimensions and metrics for evaluating recommendation systems.*" In *Recommendation Systems in Software Engineering*, pp. 245–273. Springer (2014).

[4] J. Sun, W. Du, and N. Shi, "A survey of kNN algorithm."*Information Engineering and Applied Computing*, pp. 1–10 (2018).

[5] V. Pestov, "Is the k-NN classifier in high dimensions affected by the curse of dimensionality?" *Computers & Mathematics with Applications*, 65(10) (2011).

[6] M. Kumar, D.K. Yadav, A. Singh, and V.K. Gupta, "A movie recommender system: Movrec."*International Journal of Computer Applications*, 124(3) (2015).

[7] G. Adomavicius, and A. Tuzhilin. "Toward the next generation of recommender systems: A survey of the state-of-the-art and possible extensions." *IEEE Transactions on Knowledge and Data Engineering* (2005).

[8] F. Ricci, L. Rokach, B. Shapira, and P.B. Kantor(eds.) *Recommender Systems Handbook*. Boston, MA: Springer(2010).

[9] M. Pazzani, and D. Billsus."Content-based recommendation systems." In P. Brusilovsky, A. Kobsa, and W. Nejdl (eds.) *The Adaptive Web: Methods and Strategies of Web Personalization*. LNCS, vol. 4321, Heidelberg: Springer (2007).

[10] A.K. Sahoo, and C. Pradhan. "Accuracy-assured privacy-preserving recommender system using hybrid-based deep learning method." In S.N. Mohanty, J.M. Chatterjee, S. Jain, A.A. Elngar, and P. Gupta (eds.), *Recommender System with Machine Learning and Artificial Intelligence: Practical Tools and Applications in Medical, Agricultural and Other Industries*, vol. 101 (2020), Scrivener Publishing LLC.

[11] A. K. Sahoo, C. Pradhan, and B. S. P. Mishra. "*SVD based privacy preserving recommendation model using optimized hybrid item-based collaborative filtering*." In *2019 International Conference on Communication and Signal Processing (ICCSP)*. IEEE, pp. 0294–0298 (2019).

[12] G. Guo, H. Wang, D. Bell, Y. Bi, and K. Greer. "*KNN model-based approach in classification*". In *OTM Confederated International Conferences On the Move to Meaningful Internet Systems* (pp. 986–996). Springer, Berlin, Heidelberg (2003).

[13] Z. Zhang, S. Wang, Z. Deng, and F. Chung."A fast decision algorithm of support vector machine." *Control and Decision*, 27(3), 459–463 (2012).

[14] B.M. Sarwar, G. Karypis, J.A. Konstan, and J.T. Riedl, "Application of dimensionality reduction in recommender system – a case study." In WebKDD 2000 (2000).

[15] M. Ge, C. Delgado-Battenfeld, and D. Jannach. "*Beyond accuracy: evaluating recommender systems by coverage and serendipity.*" In *Proceedings of the fourth ACM conference on Recommender systems*, ACM, pp.257–260 (2010).

[16] F. O. Isinkaye, Y. O. Folajimi, and B. A. Ojokoh. "Recommendation systems: Principles, methods and evaluation". *Egyptian Informatics Journal*, 16(3), 261–273 (2015).

[17] M. Garanayak, S. N. Mohanty, A. K. Jagadev, S. Sahoo, Recommender System Using Item Based Collaborative Filtering(CF) and K-Means, *International Journal of Knowledge-based and Intelligent Engineering Systems*, 23, 2, 93–101 (2019).

[18] M. Kunar, & S. N. Mohanty. *Location based Personalized Recommendation systems for the Tourists in India. International Journal of Business Intelligence and Data Mining* (2018) Press.

[19] M. Kunar, S. N. Mohanty, S. Sahoo, B. K. Rath, Crop Recommender System for the Farmers using Mamdani fuzzy Inference Model, *International Journal of Engineering & Technology*, 7(4.15), 277–280 (2018).

[20] H. Polat and W. Du, "*SVD-based collaborative filtering with privacy.*" In *ACM SAC'05*, pp. 791–795, ACM (2005).

[21] M. McLaughlin, and J. Herlocker. *"A collaborative filtering algorithm and evaluation metric that accurately model the user experience."* In *Proceedings of the SIGIR Conference on Research and Development in Information Retrieval* (2004).

[22] S. S. Choudhury, S. N. Mohanty, and A. K. Jagadev. Multimodal trust based recommender system with machine learning approaches for movie recommendation. *International Journal of Information Technology* (2021). https://doi.org/10.1007/s41870-020-00553-2

[23] M. Garanayak, S. Sahoo, S. N. Mohanty, A. K. Jagadev. An Automated Recommender System for Educational Institute in India. *EAI Endorsed Transactions on Scalable Information System*, 24(2), 1–13 (2020). doi. org/10.4108/eai.13-7-2018.163155

[24] B.H. Menze, B.M. Kelm, R. Masuch, and U. Himmelreich, "A comparison of Random Forest and its Gini importance with standard chemometric methods for the feature selection and classification of spectral data." IEEE (2019).

[25] R. Reshma, G. Ambikesh, and P. Santhi Thilagam. "Alleviating data sparsity and cold start in recommender systems using social behavior." IEEE (2016).

2

AN EXPERIMENTAL ANALYSIS OF COMMUNITY DETECTION ALGORITHMS ON A TEMPORALLY EVOLVING DATASET

B.S.A.S. RAJITA, MRINALINI SHUKLA, DEEPA KUMARI AND SUBHRAKANTA PANDA

BITS-Pilani, Hyderabad, Telangana, India

Contents

2.1 Introduction

Community detection is a fundamental and essential step in the Social Network Analysis (SNA) for studying the structural information of the SN topology and their various interactions with the network elements [1]. Such structural information is obtained by analyzing the interactions that happen among the nodes in the SN [2] when represented as graphs. These interactions change continuously over time, causing social communities to evolve. The evolution of communities can be attributed to the occurrence of one of the following events: *form, grow, merge, split, and dissolve.* To be able to study the occurrence of such events, the researchers must be able to detect the existing communities in the SNs. Community detection in SNs and its analysis has thus become indispensable for many applications in data mining [3], graph mining [4], machine learning [5], and data analysis [6]. Some community detection methods may be suitable for some targeted applications or datasets based on the properties of the network [7]. Hence, it is essential to know which community detection methods perform well or poorly on any dataset.

The basic task of any *Community Detection (CD)* algorithm is to detect the densely connected components (nodes) in the graphical representation of a SN, which necessitates validating the detected communities according to certain metrics [8]. The efficiency of any algorithm is known when these metrics are computed on large modern SNs with billions of edges. Many researchers [9–12] have implemented CD algorithms in some static social networks in order to detect the communities. But these community detection algorithms are also expected to perform well in dynamic (ever-growing) social networks. Hence, in this work, we propose to compare and analyze nine state-of-the-art existing algorithms, such as *Louvain* [13], *Multilevel*

[14, 15], *Fast Greedy* [16], *Label Propagation* [17], *Infomap* [18, 19], *Walktrap* [20], *Spinglass* [21], *Eigenvector* [22] and *Edge Betweenness* [23] on a dynamically large dataset, such as *DBLP*. To compare these CD algorithms, one needs a real dataset, time of execution, methodology, performance results, etc. [24]. The accuracy or correctness of the communities detected by these algorithms is computed by using Lancichinetti et al.'s [25] evaluation criteria: *Computation Time (CT), Modularity (M), and Clustering Coefficient(CC).* These metrics are defined as follows:

1. *Computation Time (CT):* The time taken by an algorithm to detect the communities is known as running time. It is used to find the time required to perform the computational process. This computation time is proportional to the input size. In our work, we have shown the time taken by each method to detect the communities in our dataset.

2. *Modularity (M):* Modularity is the measure of the degree of distribution of a network into modules (also called groups, clusters, or communities). The value of the modularity lies in the range of [-1,1]. Given a network with **m** edges, the expected number of edges between any two nodes, i and j, with degrees d_i and d_j respectively, is $\dfrac{d_i \times d_j}{2 \times m}$. So, the strength of a community, C is given as $C = \sum_{i \in C, j \in C} A_{i,j} - \dfrac{d_i \times d_j}{2 \times m}$. For a network with k communities and a total of m edges, the modularity, M, is given as $M = \sum_{i=1}^{k} \sum_{i \in C, j \in C} A_{ij} - \dfrac{d_i \times d_j}{2 \times m}$, where A_{ij} is an adjacency matrix of i rows and j columns. If i and j are connected by edges then the value is 1 else the value is 0.

3. *Clustering Coefficient (CC):* The clustering coefficient is a measure of the degree with which nodes in a graph tend to cluster together. Normally, nodes with high-density ties are more likely to form communities in most of the real-world networks. Supposing a node i has n_i neighbors and there are e_i

edges between their neighbors, then the clustering coefficient of node i is defined as

$$4.\ C_i = \begin{cases} 1, & \text{if } n_i > 1 \\ 0, & n_i = 0 \text{ or } 1 \end{cases}$$

In general, community detection methods find densely connected nodes, i.e. nodes with high cluster coefficient.

Thus, our proposed experimental analysis aims to answer the following three research questions:

RQ1 Are the existing CD algorithms suitable for evolving SNs?
RQ2 Are the LFR Benchmark parameters being suitable for comparing CD Algorithms?
RQ3 If so, which metrics are more significant for Evolving SNs?

The remaining chapter is organized as follows: Section 2.2 gives a brief overview of the working of different community detection algorithms; Section 2.3 gives the experimental setup and the results and performance analysis of the CD algorithms; and Section 2.4 concludes the work with some insights into our future work.

2.2 Background

This section gives a detailed understanding of *community detection*, the working of the *community detection algorithms*, and the *existing gaps*.

2.2.1 Community Detection

A *community* [26] is defined as a densely connected cluster of nodes in which the nodes are strongly connected. *Community detection* [27, 28] is a method to find these sets of densely connected nodes in the graphical representation of a social network (SN), based on the interaction (edges) between the nodes. These social networks can either be *static* or *dynamic*. In static SNs the entire dataset is considered to be fixed and represented in the form of a single graph. The interconnected nodes of such a SN graph are aggregated into a single point of contact without considering the temporal aspect of the interactions among nodes, i.e. the nodes are assumed to not have interaction after a certain time. This may lead to incorrect representation or loss of correct information [29] because the interaction in social networks is always active. By contrast, in dynamic SNs, the network is represented through temporal graphs, i.e. represent the interactions among nodes over a time period. These graphs represent a snapshot of the SN at different time scales i.e., daily, weekly, or monthly. More details on dynamic data representation and its processing are given in Section 2.3.

Supposing, in a *co-author* dataset, that we need to find a set of authors who are connected by their co-authorship interaction. As shown in Figure 2.1, supposing *a*1, *a*2 have together published a paper *p*1, *a*2, *a*3 published a paper *p*2, and *a*4, *a*5 published a paper *p*3. Now because of *a*2, the set of authors *a*1, *a*2, and *a*3 are co-related. So, <*a*1, *a*2, *a*3> form a community and the remaining nodes <*a*4, *a*5> form another community. But this co-author relationship of <*a*1, *a*2, *a*3> and <*a*4, *a*5> is temporal in nature, i.e. these detected communities may evolve. Such hidden relationships among the entities (community) can be known only by the application of community detection algorithms. Additionally, the performance of community detection algorithms needs to be validated through metrics (*CT*, *M*, *CC*). The communities detected (by the community detection algorithms) are compared (based on the computed metrics) to infer which algorithm would be best suited for the job (detecting communities in evolving datasets). Normally, the evolution time for different datasets would vary. In this chapter, the evolution time window considered for the co-author dataset (to represent the temporal interactions among the nodes) is one year. This is because it has been observed that it takes

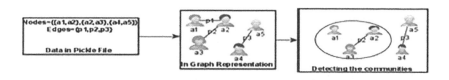

Figure 2.1 Co-authorship interaction in DBLP dataset (a) and (b): Example for Label Propagation, walktrap and spin

approximately a year for a paper to get published and appear in the *co-author* dataset. Thus, the main challenge in any evolving SN (temporal dataset) is to correctly detect the communities.

2.2.2 Community Detection Approach

Community detection is an unsupervised learning approach similar to the clustering technique in data mining. Existing community detection algorithms implement the following steps to detect the communities. First, every node is considered as a single cluster. Second, identify a set of high-density nodes as a community (also called *criterion step*). The criterion varies for each algorithm. There are three criteria-based strategies to find a set of density nodes. The first strategy is based on *topological* properties, according to this approach highly dense set of nodes can be grouped based on the *in and out count* of the edges. The second strategy is based on *structural* properties [27] such as the *number of nodes, node degree, average degree*, etc. Communities are formed based on their structural similarity. And the third strategy is based on a *community-level* approach to form a set of groups. Based on these strategies, the following community detection algorithms are selected for our implementation analysis: *Louvain* [13], *Label Propagation* [17], *Fast Greedy* [20], *Eigenvector* [30], *Infomap* [18, 19], *Multilevel* [14, 15], *Walktrap* [21], *Spinglass* [31] and *Edge Betweenness* [32]. These algorithms are implemented in Python. We have taken a very small graph (representing a synthetic dataset) consisting of 10 nodes and 14 edges to demonstrate the working of these algorithms for a better understanding. Hence, the detected communities may appear the same for some algorithms. The implementation results of these algorithms on a large dataset and its comparison is given in Section 2.3.

2.2.2.1 Label Propagation
Label Propagation [17] is a community detection algorithm for assigning labels to the unlabeled nodes. Generally, a subset of nodes (very few) is initialized with labels at the beginning of the algorithm. These labels are propagated to the unlabeled nodes. Subsequently, all the labeled nodes with similar labels form a community. Algorithm 2.1 gives the pseudo-code of Label Propagation.

ALGORITHM 2.1:   LABEL PROPAGATION

Input: A graph $G = (V,E)$, labels Y
Output: Labels $\hat{Y}$

1. *Compute Adjacency Matrix, A_{ij}*
2. *Compute Diagonal Matrix, $D_{ii} = \Sigma A_{ij}$, where D_{ii} is the degree of node i.*
3. *Compute Probability Matrix, $P = D^{-1}A_{ij}$*
4. *Initialize $Y^{(0)} = (Y^0, 0)$ at t=0, // t- first iteration*
5. *Repeat: until $Y^{(t+1)}$ converges*
6. *$Y^{(t+1)} \leftarrow P \times Y^{(t)}$*
7. *$\hat{Y} \leftarrow Y^{(t+1)}$*

In this algorithm, A_{ij} is an adjacency matrix. If there is an edge from node i to node j, then $A_{ij} = 1$, else $A_{ij} = 0$. If the probability of reaching node j from node i is v, then $P_{ij} = v$, else $P_{ij} = 0$, where $P = D^{-1}A_{ij}$ (Step 3). The next step is to assign labels to a subset of nodes and remaining unlabeled nodes are stored in $Y^{(0)}$. The assignment of labels to the unlabeled nodes can be calculated by multiplying the probability matrix (P) with the label matrix (Y). This step is repeated until nodes of Y converge. One example input for Algorithm 2.1 is shown in Figure 2.2a.

Step-1: Adjacency matrix of Figure 2.2a is:

$$
A = \begin{bmatrix}
 & 1 & 2 & 3 & 4 & 5 & 6 \\
1 & 0 & 1 & 1 & 0 & 0 & 0 \\
2 & 1 & 0 & 1 & 0 & 0 & 0 \\
3 & 1 & 1 & 0 & 1 & 0 & 0 \\
4 & 0 & 0 & 1 & 0 & 1 & 1 \\
5 & 0 & 0 & 0 & 1 & 0 & 1 \\
6 & 0 & 0 & 0 & 1 & 1 & 0
\end{bmatrix}, D = \begin{bmatrix}
 & 1 & 2 & 3 & 4 & 5 & 6 \\
1 & 2 & 0 & 0 & 0 & 0 & 0 \\
2 & 0 & 2 & 0 & 0 & 0 & 0 \\
3 & 0 & 0 & 3 & 0 & 0 & 0 \\
4 & 0 & 0 & 0 & 3 & 0 & 0 \\
5 & 0 & 0 & 0 & 0 & 2 & 0 \\
6 & 0 & 0 & 0 & 0 & 0 & 2
\end{bmatrix}
$$

Step-2: Compute the diagonal matrix, D.
Step-3: Probability matrix, $P = D^{-1}A$, is

$$P = \begin{bmatrix} 1/2 & 0 & 0 & 0 & 0 & 0 \\ 0 & 1/2 & 0 & 0 & 0 & 0 \\ 0 & 0 & 1/3 & 0 & 0 & 0 \\ 0 & 0 & 0 & 1/3 & 0 & 0 \\ 0 & 0 & 0 & 0 & 1/2 & 0 \\ 0 & 0 & 0 & 0 & 0 & 1/2 \end{bmatrix} * \begin{bmatrix} 0 & 1 & 1 & 0 & 0 & 0 \\ 1 & 0 & 1 & 0 & 0 & 0 \\ 1 & 1 & 0 & 1 & 0 & 0 \\ 0 & 0 & 1 & 0 & 1 & 1 \\ 0 & 0 & 0 & 1 & 0 & 1 \\ 0 & 0 & 0 & 1 & 1 & 0 \end{bmatrix}$$

$$= \begin{array}{c|cccccc} & 1 & 2 & 3 & 4 & 5 & 6 \\ \hline 1 & 0 & 0.5 & 0.5 & 0 & 0 & 0 \\ 2 & 0.5 & 0 & 0.5 & 0 & 0 & 0 \\ 3 & 0.3 & 0.3 & 0 & 0.3 & 0 & 0 \\ 4 & 0 & 0 & 0.3 & 0 & 0.3 & 0.3 \\ 5 & 0 & 0 & 0 & 0.5 & 0 & 0.5 \\ 6 & 0 & 0 & 0 & 0.5 & 0.5 & 0 \end{array}, \; Y^{(0)} = \begin{bmatrix} 1 \\ 0 \\ 0 \\ 0 \\ 0 \\ 0 \end{bmatrix}$$

Step-4: In this example, we assigned label 1 to node 1 (randomly picked) and the other nodes remain unlabeled (0). So, initial labeling is shown in the matrix $Y^{(0)}$.

Step-5 to Step-7: $Y^{(1)}$ is calculated using the following equation: $Y^{(t+1)} \leftarrow P \times Y^{(t)}$. So, $Y^1 \leftarrow P \times Y^0$, is represented as follows:

$$Y^{(1)} = \begin{bmatrix} 1/2 & 0 & 0 & 0 & 0 & 0 \\ 0 & 1/2 & 0 & 0 & 0 & 0 \\ 0 & 0 & 1/3 & 0 & 0 & 0 \\ 0 & 0 & 0 & 1/3 & 0 & 0 \\ 0 & 0 & 0 & 0 & 1/2 & 0 \\ 0 & 0 & 0 & 0 & 0 & 1/2 \end{bmatrix} * \begin{bmatrix} 1 \\ 0 \\ 0 \\ 0 \\ 0 \\ 0 \end{bmatrix} = \begin{bmatrix} 0 \\ 0.5 \\ 0.3 \\ 0 \\ 0 \\ 0 \end{bmatrix},$$

$$Y^{(2)} = \begin{bmatrix} 0.4 \\ 0.15 \\ 0.15 \\ 0.19 \\ 0 \\ 0 \end{bmatrix} Y^{(3)} = \begin{bmatrix} 0.215 \\ 0.275 \\ 0.219 \\ 0.045 \\ 0.045 \\ 0.045 \end{bmatrix}$$

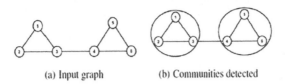

(a) Input graph　　　　(b) Communities detected

Figure 2.2　Example for Label propagation, walktrap and Spinglass

Similarly $Y^{(2)} \leftarrow P \times Y^{(1)}$ and $Y^{(3)} \leftarrow P \times Y^{(2)}$ are calculated. From $Y^{(3)}$, we can observe that nodes 1, 2, and 3 have converged to a value of 0.2, and nodes 4, 5, and 6 converged to a value of 0.045. Hence, the algorithm terminates at this point. Any further iterations beyond this point will result in some different convergence values, but the labels would remain the same for the same set of nodes. So, labels for node 1 to node 6 are {0.2,0.2,0.2,0.045,0.045, 0.045}, respectively. Therefore, the Label Propagation algorithm applied on the graph (Figure 2.2) detected two communities, C1 = {node 1, node 2, node 3} and C2 = {node 4, node 5, node 6}. This result is represented in Figure 2.2b.

2.2.2.2 Walktrap　The Walktrap algorithm [21], proposed by Pon and Latapy, detects the communities in a graph by computing the similarity distance between each pair of nodes. In this analysis similarity distance is defined as the probability of reaching a node from another node. Let $P_{i,k}$ denote the probability of reaching node k from node i. Any two nodes, node i and node j are combined to form a community, iff $P_{i,k} = P_{k,j}$. After the merger of node i and node j, they are considered as a single node, and then the computation continues. Algorithm 2.2 gives the pseudo-code of Walktrap.

ALGORITHM 2.2:　WALKTRAP

Input: A graph G = (V, E)
Output: Communities, C

1. *Compute Adjacency Matrix, A_{ij}*
2. *Compute Diagonal Matrix, $D_{ii} = \Sigma A_{ij}$ // where D_{ii} is the degree of node i.*

3. *Compute Probability Matrix $P = D^{-1}A_{ij}$*
4. *Repeat: Loop*
5. *if $p_{ik} == p_{kj}$*
6. *begin*
7. *$V_t = Merge(i,j)$*
8. *Compute A_{tij}, $D_{tii} = \Sigma A_{tij}$, $P_t = D_t^{-1}A_{tij}$*
9. *End*
10. *Else*
11. *break the Loop*
12. *for all elements of P_t*
13. *if $p_{ij} == 1$*
14. *$C_i = V_i \cup V_j$*

If there is an edge from node i to node j in the adjacency matrix, then $A_{ij} = 1$, else $A_{ij} = 0$. If the probability of reaching node j from node i is v, then $P_{ij} = v$, else $P_{ij} = 0$, where $P = D^{-1}A_{ij}$ (Step 3). The next step is to merge similar vertices based on the computed probabilities, i.e. merge node i and node j, iff $p_{i,k} = = p_{k,j}$. The merger of nodes is updated in the adjacency matrix, the diagonal matrix, and the probability matrix. This step is repeated until no similar nodes are detected to merge. Let example input for Algorithm 2.2 be the graph shown in Figure 2.2.

Step-1: Adjacency matrix of Figure 2.2a is

$$
A = \begin{array}{c|cccccc}
 & 1 & 2 & 3 & 4 & 5 & 6 \\
\hline
1 & 0 & 1 & 1 & 0 & 0 & 0 \\
2 & 1 & 0 & 1 & 0 & 0 & 0 \\
3 & 1 & 1 & 0 & 1 & 0 & 0 \\
4 & 0 & 0 & 1 & 0 & 1 & 1 \\
5 & 0 & 0 & 0 & 1 & 0 & 1 \\
6 & 0 & 0 & 0 & 1 & 1 & 0 \\
\end{array}
, \quad
D = \begin{array}{c|cccccc}
 & 1 & 2 & 3 & 4 & 5 & 6 \\
\hline
1 & 2 & 0 & 0 & 0 & 0 & 0 \\
2 & 0 & 2 & 0 & 0 & 0 & 0 \\
3 & 0 & 0 & 3 & 0 & 0 & 0 \\
4 & 0 & 0 & 0 & 3 & 0 & 0 \\
5 & 0 & 0 & 0 & 0 & 2 & 0 \\
6 & 0 & 0 & 0 & 0 & 0 & 2 \\
\end{array}
$$

Step-2: Compute the diagonal matrix, D.
Step-3: Probability matrix, $P = D^{-1}A$, is

$$P = \begin{bmatrix} 1/2 & 0 & 0 & 0 & 0 & 0 \\ 0 & 1/2 & 0 & 0 & 0 & 0 \\ 0 & 0 & 1/3 & 0 & 0 & 0 \\ 0 & 0 & 0 & 1/3 & 0 & 0 \\ 0 & 0 & 0 & 0 & 1/2 & 0 \\ 0 & 0 & 0 & 0 & 0 & 1/2 \end{bmatrix} * \begin{bmatrix} 0 & 1 & 1 & 0 & 0 & 0 \\ 1 & 0 & 1 & 0 & 0 & 0 \\ 1 & 1 & 0 & 1 & 0 & 0 \\ 0 & 0 & 1 & 0 & 1 & 1 \\ 0 & 0 & 0 & 1 & 0 & 1 \\ 0 & 0 & 0 & 1 & 1 & 0 \end{bmatrix}$$

$$= \begin{array}{c|cccccc} & 1 & 2 & 3 & 4 & 5 & 6 \\ \hline 1 & 0 & 0.5 & 0.5 & 0 & 0 & 0 \\ 2 & 0.5 & 0 & 0.5 & 0 & 0 & 0 \\ 3 & 0.3 & 0.3 & 0 & 0.3 & 0 & 0 \\ 4 & 0 & 0 & 0.3 & 0 & 0.3 & 0.3 \\ 5 & 0 & 0 & 0 & 0.5 & 0 & 0.5 \\ 6 & 0 & 0 & 0 & 0.5 & 0.5 & 0 \end{array}$$

Step-4: Find similar nodes in matrix P. Since, $P_{1,2} == P_{2,3} == 0.5$ implies node 1 and node 3 are similar, hence these nodes are merged. For similar reasons, node 4 and node 5 are also merged. Now, the newly updated adjacency matrix, diagonal matrix, and probability matrix are shown below:

$$A_1 = \begin{array}{c|cccc} & (1,3) & 2 & (4,5) & 6 \\ \hline (1,3) & 0 & 2 & 0 & 0 \\ 2 & 2 & 0 & 1 & 0 \\ (4,5) & 0 & 1 & 0 & 2 \\ 6 & 0 & 0 & 2 & 0 \end{array}, D_1 = \begin{array}{c|cccc} & (1,3) & 2 & (4,5) & 6 \\ \hline (1,3) & 2 & 0 & 0 & 0 \\ 2 & 0 & 0 & 3 & 0 \\ (4,5) & 0 & 0 & 3 & 0 \\ 6 & 0 & 0 & 0 & 2 \end{array}$$

$$P_1 = \begin{bmatrix} 1/2 & 0 & 0 & 0 \\ 0 & 1/3 & 0 & 0 \\ 0 & 0 & 1/3 & 0 \\ 0 & 0 & 0 & 1/2 \end{bmatrix} * \begin{bmatrix} 0 & 2 & 0 & 0 \\ 2 & 0 & 1 & 0 \\ 0 & 1 & 0 & 2 \\ 0 & 0 & 2 & 0 \end{bmatrix} = \begin{bmatrix} 0 & 1 & 0 & 0 \\ 0.6 & 0 & 0.3 & 0 \\ 0 & 0.3 & 0 & 0.6 \\ 0 & 0 & 1 & 0 \end{bmatrix}$$

Since, no similar vertices are found; hence, the algorithm terminates at this point.

Step-5: Form the communities. To form the communities, we need to find and merge those nodes in the probability matrix whose corresponding value equals 1. Therefore, Walktrap algorithm applied on the graph (Figure 2.2) detected two communities, C1 = {node 1, node 3, node 2} and C2 = {node 4, node 5, node 6}, as shown encircled in Figure 2.2b.

2.2.2.3 Spinglass The Spinglass [31] algorithm uses a random walk to compute the probability of reaching from one node to another in a graph. It computes the Hamiltonian distance between the nodes based on the probability value. This measurement gives the nodes having a similar spin state. So, this algorithm groups those nodes which have similar Hamiltonian distances. This grouping process is repeated until no similar nodes exist. The pseudo-code of Spinglass is shown in Algorithm 2.3.

If there is an edge from node i to node j in the adjacency matrix, then $A_{ij} = 1$, else $A_{ij} = 0$. The next step is to find the Hamiltonian Matrix. Any two nodes, i and j in H, having the highest value are considered to be in the same spin state and hence are merged to

---

ALGORITHM 2.3:   SPINGLASS

Input: A graph G = (V,E) labels Y
Output: Communities C

1. *Compute Adjacency Matrix, A_{ij}*
2. *Compute Diagonal Matrix, $D_{ii} = \Sigma A_{ij}$ // where D_{ii} is the degree of node i.*
3. *Compute Probability Matrix, $P = D^{-1}A_{ij}$*
4. *Compute Hamiltonian Matrix, $H = A_{ij} - P$*
5. *Repeat:*
6. *find max_elementt(H)*
7. *$V_t = Merge(i,j)$*
8. *Compute Adjacency Matrix, Diagonal Matrix, $D_{tii} = \Sigma A_{tij}$, Probability Matrix, $P_t = D_t^{-1}A_{t_{ij}}$, Hamiltonian Matrix, $H_t = A_{tij} - P_t$*
9. *Until $H_t == [0]$*
10. *for all elements of D_t*
11. *if $D_{tij} == 1$*
12. *$C_i = V_i \cup V_j$*

form a single node. This is updated in the adjacency matrix, diagonal matrix, probability matrix, and Hamiltonian matrix. This process is repeated until the Hamiltonian Matrix is null. Let the same graph in Figure 2.2a be the example input for Algorithm 2.3.

Step-1: Adjacency matrix of the graph in Figure 2.2a is

$$
A = \begin{array}{c|cccccc}
 & 1 & 2 & 3 & 4 & 5 & 6 \\
\hline
1 & 0 & 1 & 1 & 0 & 0 & 0 \\
2 & 1 & 0 & 1 & 0 & 0 & 0 \\
3 & 1 & 1 & 0 & 1 & 0 & 0 \\
4 & 0 & 0 & 1 & 0 & 1 & 1 \\
5 & 0 & 0 & 0 & 1 & 0 & 1 \\
6 & 0 & 0 & 0 & 1 & 1 & 0
\end{array}
, D = \begin{array}{c|cccccc}
 & 1 & 2 & 3 & 4 & 5 & 6 \\
\hline
1 & 2 & 0 & 0 & 0 & 0 & 0 \\
2 & 0 & 2 & 0 & 0 & 0 & 0 \\
3 & 0 & 0 & 3 & 0 & 0 & 0 \\
4 & 0 & 0 & 0 & 3 & 0 & 0 \\
5 & 0 & 0 & 0 & 0 & 2 & 0 \\
6 & 0 & 0 & 0 & 0 & 0 & 2
\end{array}
$$

Step-2: Compute the diagonal matrix, D.
Step-3: Probability matrix, $P = D^{-1}A$, is

$$
P = \begin{bmatrix}
1/2 & 0 & 0 & 0 & 0 & 0 \\
0 & 1/2 & 0 & 0 & 0 & 0 \\
0 & 0 & 1/3 & 0 & 0 & 0 \\
0 & 0 & 0 & 1/3 & 0 & 0 \\
0 & 0 & 0 & 0 & 1/2 & 0 \\
0 & 0 & 0 & 0 & 0 & 1/2
\end{bmatrix}
*
\begin{bmatrix}
0 & 1 & 1 & 0 & 0 & 0 \\
1 & 0 & 1 & 0 & 0 & 0 \\
1 & 1 & 0 & 1 & 0 & 0 \\
0 & 0 & 1 & 0 & 1 & 1 \\
0 & 0 & 0 & 1 & 0 & 1 \\
0 & 0 & 0 & 1 & 1 & 0
\end{bmatrix}
$$

$$
= \begin{array}{c|cccccc}
 & 1 & 2 & 3 & 4 & 5 & 6 \\
\hline
1 & 0 & 0.5 & 0.5 & 0 & 0 & 0 \\
2 & 0.5 & 0 & 0.5 & 0 & 0 & 0 \\
3 & 0.3 & 0.3 & 0 & 0.3 & 0 & 0 \\
4 & 0 & 0 & 0.3 & 0 & 0.3 & 0.3 \\
5 & 0 & 0 & 0 & 0.5 & 0 & 0.5 \\
6 & 0 & 0 & 0 & 0.5 & 0.5 & 0
\end{array}
$$

Step-4: Hamiltonian matrix, $H = A_{ij} - P$.

$$H = \begin{bmatrix} 0 & 1 & 1 & 0 & 0 & 0 \\ 1 & 0 & 1 & 0 & 0 & 0 \\ 1 & 1 & 0 & 1 & 0 & 0 \\ 0 & 0 & 1 & 0 & 1 & 1 \\ 0 & 0 & 0 & 1 & 0 & 1 \\ 0 & 0 & 0 & 1 & 1 & 0 \end{bmatrix} - \begin{bmatrix} 0 & 0.5 & 0.5 & 0 & 0 & 0 \\ 0.5 & 0 & 0.5 & 0 & 0 & 0 \\ 0.3 & 0.3 & 0 & 0.3 & 0 & 0 \\ 0 & 0 & 0.3 & 0 & 0.3 & 0.3 \\ 0 & 0 & 0 & 0.5 & 0 & 0.5 \\ 0 & 0 & 0 & 0.5 & 0.5 & 0 \end{bmatrix}$$

	1	2	3	4	5	6
1	0	0.5	0.5	0	0	0
2	0.5	0	0.5	0	0	0
= 3	0.7	0.7	0	0.7	0	0
4	0	0	0.7	0	0.7	0.7
5	0	0	0	0.5	0	0.5
6	0	0	0	0.5	0.5	0

Step-5: Similar spin states can be identified by finding the maximum value between any pair of nodes, i and j. In the Hamiltonian matrix, H, node 3 and node 1 have the highest spin state, as H_{31} value is the highest in H. So, node 1 and node 3 are merged. Now, the newly updated adjacency matrix, the diagonal matrix, the probability matrix, and the Hamiltonian matrix are shown below:

$$A_1 \begin{array}{c|ccccc} & (1,3) & 2 & 4 & 5 & 6 \\ \hline (1,3) & 0 & 2 & 0 & 0 & 0 \\ 2 & 2 & 0 & 1 & 0 & 0 \\ 4 & 0 & 1 & 0 & 2 & 1 \\ 5 & 0 & 0 & 2 & 0 & 1 \\ 6 & 0 & 0 & 1 & 1 & 0 \end{array} , D_1 \begin{array}{c|ccccc} & (1,3) & 2 & 4 & 5 & 6 \\ \hline (1,3) & 0.5 & 0 & 0 & 0 & 0 \\ 2 & 0 & 0.3 & 0 & 0 & 0 \\ 4 & 0 & 0 & 0.3 & 0 & 0 \\ 5 & 0 & 0 & 0 & 0.5 & 0 \\ 6 & 0 & 0 & 0 & 0 & 0.5 \end{array} ,$$

$$P_1 \begin{array}{c|ccccc} & (1,3) & 2 & 4 & 5 & 6 \\ \hline (1,3) & 0 & 1 & 0 & 0 & 0 \\ 2 & 0.6 & 0 & 0.3 & 0 & 0 \\ 4 & 0 & 0.3 & 0 & 0.3 & 0.3 \\ 5 & 0 & 0 & 0.5 & 0 & 0.5 \\ 6 & 0 & 0 & 0.5 & 0.5 & 0 \end{array}$$

$$H_1 = \begin{bmatrix} 0 & 2 & 0 & 0 & 0 \\ 2 & 0 & 1 & 0 & 0 \\ 0 & 1 & 0 & 2 & 1 \\ 0 & 0 & 2 & 0 & 1 \\ 0 & 0 & 1 & 1 & 0 \end{bmatrix} - \begin{bmatrix} 0 & 1 & 0 & 0 & 0 \\ 0.6 & 0 & 0.3 & 0 & 0 \\ 0 & 0.3 & 0 & 0.3 & 0.3 \\ 0 & 0 & 0.5 & 0 & 0.5 \\ 0 & 0 & 0.5 & 0.5 & 0 \end{bmatrix}$$

$$= \begin{bmatrix} 0 & 1 & 0 & 0 & 0 \\ 1.4 & 0 & 0.7 & 0 & 0 \\ 0 & 0.7 & 0 & 0.7 & 0.7 \\ 0 & 0.7 & 0 & 0.7 & 0.7 \\ 0 & 0 & 0.5 & 0 & 0.5 \\ 0 & 0 & 0.5 & 0.5 & 0 \end{bmatrix}$$

Step-6: Similar spin (highest spin state) in H_1 is found between a node (1,3) and node 2. Step-7: Node (1,3) and node 2 are merged to form (1,3,2). This process continues. At Step-8, the Hamiltonian matrix becomes null and hence the iteration stops. The final state of the matrices and computation is shown below:

$$A_4 = \begin{bmatrix} & (1,3,2) & (4,5,6) \\ (1,3,2) & 1 & 1 \\ (4,5,6) & 1 & 1 \end{bmatrix}, D_4^{-1} = \begin{bmatrix} & (1,3,2) & (4,5,6) \\ (1,3,2) & 1 & 0 \\ (4,5,6) & 0 & 1 \end{bmatrix}$$

$$P_4 = \begin{bmatrix} & (1,3,2) & (4,5,6) \\ (1,3,2) & 1 & 1 \\ (4,5,6) & 1 & 1 \end{bmatrix} H_4 = \begin{bmatrix} 1 & 1 \\ 1 & 1 \end{bmatrix} - \begin{bmatrix} 1 & 1 \\ 1 & 1 \end{bmatrix}$$

$$= \begin{bmatrix} & (1,3,2) & (4,5,6) \\ (1,3,2) & 0 & 0 \\ (4,5,6) & 0 & 0 \end{bmatrix}$$

Step-9: To find the final set of communities, we look into the diagonal matrix for those set of nodes whose value equals 1, and merge them. Therefore, the Spinglass algorithm applied on the graph detected two communities, C1 = {node 1, node 2, node 3} and C2 = {node 4, node 5, node 6} as shown encircled in Figure 2.2a.

2.2.2.4 Infomap The Infomap [18, 19] algorithm assigns Huffman code to all nodes of the input graph in order to detect the communities. The random walk is used to identify all those nodes with the same Huffman code and it then groups these nodes together to form a community. This process is repeated until no nodes with the same Huffman code exist. The pseudo-code of Infomap is shown in Algorithm 2.4.

If there is an edge from node i to node j in the adjacency matrix, then $A_{ij} = 1$, else $A_{ij} = 0$. If the probability of reaching node j from node i is v, then $P_{ij} = v$, else $P_{ij} = 0$, where $P = D^{-1}A_{ij}$ (Step 3). The next step is to assign Huffman code values to all nodes of the graph and is stored in $H^{(0)}$. The assigned labels of the nodes can be calculated by

---

ALGORITHM 2.4:   INFOMAP

Input: A graph G = (V, E)
Output: Labels $\hat{Y}$

1. *Compute Adjacency Matrix, A_{ij}*
2. *Compute Diagonal Matrix, $D_{ii} = \Sigma A_{ij}$, where D_{ii} is the degree of node i.*
3. *Compute Probability Matrix, $P = D^{-1}A_{ij}$*
4. *Initialize $\forall V \in G$, $H^{(t)} = Huffman\_code$ at $t = 0$, // implying first iteration*
5. *Initialize $Y^{(0)} \leftarrow P \times H^{(0)}$,*
6. *Repeat: until $Y^{(t)}$ converges*
7. *if($Y_i^{(t)} == Y_j^{(t)}$)*
8. *$H_i^{(t+1)} \leftarrow H_j^{(t)}$*
9. *$Y^{(t+1)} \leftarrow P \times H^{(t+1)}$*
10. *$Y \leftarrow Y^{(t+1)}$*

multiplying the probability matrix (P) with the Huffman code matrix (H) as shown in Step 5. Then assign the same Huffman code value to the nodes, if those nodes have similar labels. This step is repeated until nodes of Y converge. One example input for Algorithm 2.4 is shown in Figure 2.3.

Step-1: Adjacency matrix and Diagonal matrix for Figure 2.3a are

$$
A =
\begin{array}{c|cccccccccc}
 & 1 & 2 & 3 & 4 & 5 & 6 & 7 & 8 & 9 & 10 \\
\hline
1 & 0 & 1 & 1 & 0 & 1 & 0 & 0 & 0 & 0 & 0 \\
2 & 1 & 0 & 1 & 1 & 1 & 0 & 0 & 1 & 0 & 0 \\
3 & 1 & 1 & 0 & 0 & 1 & 0 & 1 & 0 & 0 & 0 \\
4 & 0 & 1 & 0 & 0 & 0 & 1 & 1 & 0 & 0 & 0 \\
5 & 1 & 1 & 1 & 0 & 0 & 0 & 0 & 0 & 0 & 0 \\
6 & 0 & 0 & 0 & 0 & 1 & 0 & 0 & 1 & 0 & 0 \\
7 & 0 & 0 & 1 & 1 & 0 & 1 & 0 & 0 & 0 & 0 \\
8 & 0 & 1 & 0 & 0 & 0 & 0 & 0 & 0 & 1 & 1 \\
9 & 0 & 0 & 0 & 0 & 0 & 0 & 0 & 1 & 0 & 1 \\
10 & 0 & 0 & 0 & 0 & 0 & 1 & 0 & 1 & 1 & 0 \\
\end{array}
$$

$$
D =
\begin{array}{c|cccccccccc}
 & 1 & 2 & 3 & 4 & 5 & 6 & 7 & 8 & 9 & 10 \\
\hline
1 & 3 & 0 & 0 & 0 & 0 & 0 & 0 & 0 & 0 & 0 \\
2 & 0 & 4 & 0 & 0 & 0 & 0 & 0 & 0 & 0 & 0 \\
3 & 0 & 0 & 4 & 0 & 0 & 0 & 0 & 0 & 0 & 0 \\
4 & 0 & 0 & 0 & 3 & 0 & 0 & 0 & 0 & 0 & 0 \\
5 & 0 & 0 & 0 & 0 & 3 & 0 & 0 & 0 & 0 & 0 \\
6 & 0 & 0 & 0 & 0 & 0 & 3 & 0 & 0 & 0 & 0 \\
7 & 0 & 0 & 0 & 0 & 0 & 0 & 3 & 0 & 0 & 0 \\
8 & 0 & 0 & 0 & 0 & 0 & 0 & 0 & 3 & 0 & 0 \\
9 & 0 & 0 & 0 & 0 & 0 & 0 & 0 & 0 & 2 & 0 \\
10 & 0 & 0 & 0 & 0 & 0 & 3 & 0 & 0 & 0 & 3 \\
\end{array}
$$

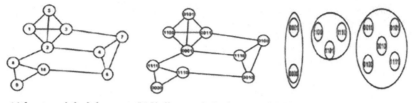

(a) Input graph for Info map   (b) Huffman code for Input graph (c) Communities detected by Info for Infomap   map

Figure 2.3   (a)-(c): Infomap Input graph and Huffman coded graph

$$
P = \begin{array}{c|cccccccccc}
 & 1 & 2 & 3 & 4 & 5 & 6 & 7 & 8 & 9 & 10 \\
\hline
1 & 0 & 0.3 & 0.3 & 0 & 0.3 & 0 & 0 & 0 & 0 & 0 \\
2 & 0.25 & 0 & 0.25 & 0.25 & 0.25 & 0 & 0 & 0.25 & 0 & 0 \\
3 & 0.25 & 0.25 & 0 & 0 & 0.25 & 0 & 0.25 & 0 & 0 & 0 \\
4 & 0 & 0.3 & 0 & 0 & 0 & 0.3 & 0.3 & 0 & 0 & 0 \\
5 & 0.3 & 0.3 & 0.3 & 0 & 0 & 0 & 0 & 0 & 0 & 0 \\
6 & 0 & 0 & 0 & 0 & 0.3 & 0 & 0 & 0.3 & 0 & 0 \\
7 & 0 & 0 & 0.3 & 0.3 & 0 & 0.3 & 0 & 0 & 0 & 0 \\
8 & 0 & 0.3 & 0 & 0 & 0 & 0 & 0 & 0 & 0.3 & 0.3 \\
9 & 0 & 0 & 0 & 0 & 0 & 0 & 0 & 0.5 & 0 & 0.5 \\
10 & 0 & 0 & 0 & 0 & 0 & 0.3 & 0 & 0.3 & 0.3 & 0 \\
\end{array}
$$

Step-2: Compute probability matrix, $P = D^{-1}A$.

Step-3: In this example, we assign Huffman code to all nodes as follows (the graph is shown in Figure 2.3).

$$
H^{(0)} = \begin{bmatrix}
1 - 1100 \\
2 - 0001 \\
3 - 0011 \\
4 - 1110 \\
5 - 0101 \\
6 - 0010 \\
7 - 0100 \\
8 - 1111 \\
9 - 0000 \\
10 - 1110
\end{bmatrix},
Y^{(0)} = \begin{bmatrix}
1 - 33.9 \\
2 - 858.25 \\
3 - 325.5 \\
4 - 33.3 \\
5 - 333.6 \\
6 - 69.6 \\
7 - 339.3 \\
8 - 333.3 \\
9 - 555 \\
10 - 336.3
\end{bmatrix},
Y^{(3)} = \begin{bmatrix}
1 - 60.9 \\
2 - 625.25 \\
3 - 325.75 \\
4 - 60.9 \\
5 - 325.75 \\
6 - 325.75 \\
7 - 325.75 \\
8 - 325.75 \\
9 - 625.25 \\
10 - 60.9
\end{bmatrix},
H^{(3)} = \begin{bmatrix}
1 - 1100 \\
2 - 0001 \\
3 - 0011 \\
4 - 1100 \\
5 - 0011 \\
6 - 0011 \\
7 - 0011 \\
8 - 0011 \\
9 - 0001 \\
10 - 1100
\end{bmatrix}
$$

Step-4: In this step, the algorithm calculates labels of the nodes using, $Y^{(0)} = P \times H^{(0)}$.

Step-5: Assign the same Huffman code to those set of nodes, which have similar label values (nodes with labels having the same integral value) in $Y^{(0)}$. For example, node 1 and node 4 are considered to have similar labels. So, node 1 and node 4 will get the same Huffman code. This process continues. At Step-9, the Huffman matrix converges and hence the iteration stops. The final state of the matrices are shown $H^{(3)}$.

From $Y^{(3)}$, we can observe that nodes 1, 4, and 10 have converged to a value of 60.9, nodes 2 and 9 have converged to a value of 625.75, and nodes 3,5,6,7, and 8 converged to a value of 325.75. Hence, the algorithm terminates at this point. Therefore, the Infomap algorithm applied on the graph (Figure 2.3) detected the following three communities: C1 = {node 1, node 4, node 10}, C2 = {node 2, node 9}, and C3 = {node 3, node 5, node 6, node 7, node 8}. This result is shown in Figure 2.3c. From the result, it can be inferred that the Huffman code of every node in a community is the same. Hence, all the nodes having the same Huffman code can be grouped to form a community.

2.2.2.5 Eigenvector The Eigenvector [30] algorithm, also called *Spectral clustering*, was proposed by Newman. In this approach, the modularity of the adjacency matrix of a given graph is represented in the form of eigenvalues. For each node in the graph, the 1st degree Eigenvalues are computed. All those nodes with similar 1st degree eigenvalues are grouped together to form the communities. Algorithm 2.5 gives the pseudo-code of the Leading Eigenvector.

If there is an edge from node i to node j in the adjacency matrix, then $A_{ij} = 1$, else $A_{ij} = 0$. Next, it finds the Laplacian matrix using $L = D - A$. Then, 1st degree of eigenvalues is computed and stored in Spectral matrix, S. Nodes with similar 1st degree eigenvalues are grouped to get the communities. One example input for Algorithm 2.5 is shown in Figure 2.4.

ALGORITHM 2.5:　EIGENVECTOR

Input: A graph G = (V, E)
Output: C, community sets

1. Compute Adjacency Matrix, A_{ij}
2. Compute Diagonal Matrix, $D_{ii} = \Sigma A_{ij}$ // where D_{ii} is the degree of node i.
3. Compute Laplacian Matrix, L = D - A
4. Compute Spectral Matrix, S = 1st deg [L]
5. Group similar set of eigenvalues as one community.

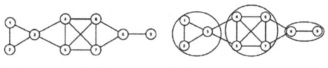

(a) Input graph for Eigenvector　(b) Communities detected by Eigenvector

Figure 2.4　a) and (b): Input graph for Eigenvector

Step-1: Adjacency matrix of Figure 2.4a is:

$$A = \begin{matrix}
 & 1 & 2 & 3 & 4 & 5 & 6 & 7 & 8 & 9 \\
1 & 0 & 1 & 1 & 0 & 0 & 0 & 0 & 0 & 0 \\
2 & 1 & 0 & 1 & 0 & 0 & 0 & 0 & 0 & 0 \\
3 & 1 & 1 & 0 & 1 & 1 & 0 & 0 & 0 & 0 \\
4 & 0 & 0 & 1 & 0 & 1 & 1 & 1 & 0 & 0 \\
5 & 0 & 0 & 1 & 1 & 0 & 1 & 1 & 0 & 0 \\
6 & 0 & 0 & 0 & 1 & 1 & 0 & 1 & 1 & 0 \\
7 & 0 & 0 & 0 & 1 & 1 & 1 & 0 & 1 & 0 \\
8 & 0 & 0 & 0 & 0 & 0 & 1 & 1 & 0 & 1 \\
9 & 0 & 0 & 0 & 0 & 0 & 0 & 0 & 1 & 0 \\
\end{matrix},$$

$$D = \begin{array}{c|ccccccccc} & 1 & 2 & 3 & 4 & 5 & 6 & 7 & 8 & 9 \\ \hline 1 & 2 & 0 & 0 & 0 & 0 & 0 & 0 & 0 & 0 \\ 2 & 0 & 2 & 0 & 0 & 0 & 0 & 0 & 0 & 0 \\ 3 & 0 & 0 & 4 & 0 & 0 & 0 & 0 & 0 & 0 \\ 4 & 0 & 0 & 0 & 4 & 0 & 0 & 0 & 0 & 0 \\ 5 & 0 & 0 & 0 & 0 & 4 & 0 & 0 & 0 & 0 \\ 6 & 0 & 0 & 0 & 0 & 0 & 4 & 0 & 0 & 0 \\ 7 & 0 & 0 & 0 & 0 & 0 & 0 & 4 & 0 & 0 \\ 8 & 0 & 0 & 0 & 0 & 0 & 0 & 0 & 3 & 0 \\ 9 & 0 & 0 & 0 & 0 & 0 & 0 & 0 & 0 & 1 \end{array}$$

Step-2: The diagonal matrix, D, is computed.

Step-3: Laplacian matrix, $L = D - A$ is:

$$L = \begin{bmatrix} 2 & 0 & 0 & 0 & 0 & 0 & 0 & 0 & 0 \\ 0 & 2 & 0 & 0 & 0 & 0 & 0 & 0 & 0 \\ 0 & 0 & 4 & 0 & 0 & 0 & 0 & 0 & 0 \\ 0 & 0 & 0 & 4 & 0 & 0 & 0 & 0 & 0 \\ 0 & 0 & 0 & 0 & 4 & 0 & 0 & 0 & 0 \\ 0 & 0 & 0 & 0 & 0 & 4 & 0 & 0 & 0 \\ 0 & 0 & 0 & 0 & 0 & 0 & 4 & 0 & 0 \\ 0 & 0 & 0 & 0 & 0 & 0 & 0 & 3 & 0 \\ 0 & 0 & 0 & 0 & 0 & 0 & 0 & 0 & 1 \end{bmatrix} - \begin{bmatrix} 0 & 1 & 1 & 0 & 0 & 0 & 0 & 0 & 0 \\ 1 & 0 & 1 & 0 & 0 & 0 & 0 & 0 & 0 \\ 1 & 1 & 0 & 1 & 1 & 0 & 0 & 0 & 0 \\ 0 & 0 & 1 & 0 & 1 & 1 & 1 & 0 & 0 \\ 0 & 0 & 1 & 1 & 0 & 1 & 1 & 0 & 0 \\ 0 & 0 & 0 & 1 & 1 & 0 & 1 & 1 & 0 \\ 0 & 0 & 0 & 1 & 1 & 1 & 0 & 1 & 0 \\ 0 & 0 & 0 & 0 & 0 & 1 & 1 & 0 & 1 \\ 0 & 0 & 0 & 0 & 0 & 0 & 0 & 1 & 0 \end{bmatrix}$$

$$= \begin{array}{c|ccccccccc} & 1 & 2 & 3 & 4 & 5 & 6 & 7 & 8 & 9 \\ \hline 1 & 2 & -1 & -1 & 0 & 0 & 0 & 0 & 0 & 0 \\ 2 & -1 & 2 & -1 & 0 & 0 & 0 & 0 & 0 & 0 \\ 3 & -1 & -1 & 4 & -1 & -1 & 0 & 0 & 0 & 0 \\ 4 & 0 & 0 & -1 & 4 & -1 & -1 & -1 & 0 & 0 \\ 5 & 0 & 0 & -1 & -1 & 4 & -1 & -1 & 0 & 0 \\ 6 & 0 & 0 & 0 & -1 & -1 & 4 & -1 & -1 & 0 \\ 7 & 0 & 0 & 0 & -1 & -1 & -1 & 4 & -1 & 0 \\ 8 & 0 & 0 & 0 & 0 & 0 & -1 & -1 & 3 & -1 \\ 9 & 0 & 0 & 0 & 0 & 0 & 0 & 0 & -1 & 0 \end{array}$$

Step-4: Compute 1st degree eigenvalues of $[L]$, where λ_i corresponds to the 1st degree eigenvalue of node i.

$$S = \begin{bmatrix} node-1 : \lambda_1 \to 5.76891 + 0.514177i \\ node-2 : \lambda_2 \to 5.76891 - 0.514177i \\ node-3 : \lambda_3 \to 5.0 \\ node-4 : \lambda_4 \to 3.54548 \\ node-5 : \lambda_5 \to 3.4648 \\ node-6 : \lambda_6 \to 3.76121 \\ node-7 : \lambda_7 \to 3.674619 \\ node-8 : \lambda_8 \to 0.474631 \\ node-9 : \lambda_9 \to 0.0445074 \end{bmatrix}$$

Step-5: Similar eigenvalue nodes (i.e. nodes whose integral eigenvalues are same) in matrix, **S**, are grouped to form the communities. Therefore, the eigenvector algorithm, applied on the graph (Figure 2.4), detects three communities as follows: C1 = {node 1, node 2, node 3}; C2 = {node 4, node 5, node 6, node 7}; and C3 = {node 8, node 9} as shown in Figure 2.4b.

2.2.2.6 Edge Betweenness The Edge Betweenness [32] algorithm is designed by Girvan and Newman. It detects the communities based on the edge betweenness measure of the nodes in a given graph. Edge betweenness is the number of paths that include a given edge. An edge is said to be densely connected if it has the highest edge betweenness. The goal of this algorithm is to remove the edges with the highest edge betweenness score. This process is repeated until the edge betweenness of each edge is 1. Algorithm 2.6 gives the pseudocode of Edge betweenness.

The edge betweenness score of all existing edges in the graph is calculated and stored in *M*. One example input for Algorithm 2.6 is shown in Figure 2.5.

Step-1: Compute edge betweenness of an edge (a, b) = number of paths in the graph of Figure 2.5 that includes the edge (a, b). Edge_betweenness(1, 3) = Number of paths connected to 1 (Node 1 is having 1 connection with node 2) x Number of paths connected to 3 (Node 3 is having connection to

ALGORITHM 2.6:   EDGE BETWEENNESS

Input : *A graph $G = (V, E)$, where $E = \{(i, j) \mid i, j \in V\}$*
Output: *C, Community sets*

1. *Compute $M_{ij} \leftarrow Edge\_Betweenness,\ \forall(i,j) \in G$*
2. *Repeat Loop until $Edge\_Betweenness(M_{ij}==1)\ \forall(i,j)$*
3. *$E' = E - \{(i,j)\}$, where $M_{ij} = m$*
4. *xconstruct $G = G'$, where $G' = (V,E')$*
5. *for all elements of M*
6. *if $M_{ij} == 1$*
7. *$C_k = V_i \cup V_p$, where $1 \le k \le |V|$*

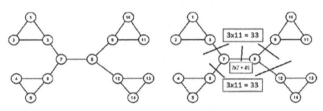

(a) Input for Edge betweenness   (b) The edges with the highest
betweenness

Figure 2.5   (a) and (b): Input graph for Edge Betweenness

remaining 12 nodes) $=1 \times 12 = 12$. Similarly remaining connected edge betweenness has to be calculated.

Step-2: Find the highest edge betweenness score. The computed
edge betweenness values for the edges of the example graph
are 49, 33, and 12. So, in this step, the edge (7, 8) has the maximum edge betweenness value of 49, hence it is removed. After
removing (7,8), the new graph, G', is shown in Figure 2.6a.

Step-3: So, again edge_betweenness values are calculated as follows: Edge_betweenness (1, 3) = Edge_betweenness
(2, 3) = Edge_betweenness (4, 6) = Edge_betweenness
(5, 6) = Edge_betweenness (9, 10) = Edge_betweenness
(10, 11) = Edge_betweenness (12, 13) = Edge_betweenness
(13, 14)$=1 \times 5 = 5$. It is shown in M'.

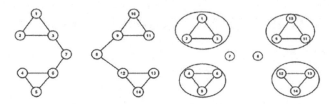

(a) The edge with the highest   (b) Output of edg betweenness
betweenness is removed

Figure 2.6   (a) and (b): Example for Edge Betweenness

Step-4: Finding the highest edge betweenness values out of 12
and 5. Now, the edges (3,7), (6,7), (8,9), and (8,12), having an
edge betweenness value of 12, are removed. After removing
the above-mentioned edges, the resultant graph is shown in
Figure 2.6. So this edge removal process continues until the
Edge_betweenness value of every edge is equal to 1. The final
matrix is given as M_{final}.

$$M' = \begin{array}{c|cccccccccccccc} & 1 & 2 & 3 & 4 & 5 & 6 & 7 & 8 & 9 & 10 & 11 & 12 & 13 & 14 \\ \hline 1 & 0 & 5 & 5 & 0 & 0 & 0 & 0 & 0 & 0 & 0 & 0 & 0 & 0 & 0 \\ 2 & 5 & 0 & 5 & 0 & 0 & 0 & 0 & 0 & 0 & 0 & 0 & 0 & 0 & 0 \\ 3 & 5 & 5 & 0 & 0 & 0 & 0 & 12 & 0 & 0 & 0 & 0 & 0 & 0 & 0 \\ 4 & 0 & 0 & 0 & 0 & 5 & 5 & 0 & 0 & 0 & 0 & 0 & 0 & 0 & 0 \\ 5 & 0 & 0 & 0 & 5 & 5 & 0 & 0 & 0 & 0 & 0 & 0 & 0 & 0 & 0 \\ 6 & 0 & 0 & 0 & 5 & 5 & 0 & 12 & 0 & 0 & 0 & 0 & 0 & 0 & 0 \\ 7 & 0 & 0 & 12 & 0 & 0 & 12 & 0 & 0 & 0 & 0 & 0 & 0 & 0 & 0 \\ 8 & 0 & 0 & 0 & 0 & 0 & 0 & 0 & 12 & 0 & 0 & 0 & 12 & 0 & 0 \\ 9 & 0 & 0 & 0 & 0 & 0 & 0 & 0 & 12 & 0 & 5 & 5 & 0 & 0 & 0 \\ 10 & 0 & 0 & 0 & 0 & 0 & 0 & 0 & 0 & 5 & 0 & 5 & 0 & 0 & 0 \\ 11 & 0 & 0 & 0 & 0 & 0 & 0 & 0 & 0 & 0 & 5 & 5 & 0 & 0 & 0 \\ 12 & 0 & 0 & 0 & 0 & 0 & 0 & 0 & 12 & 0 & 0 & 0 & 0 & 5 & 5 \\ 13 & 0 & 0 & 0 & 0 & 0 & 0 & 0 & 0 & 0 & 0 & 0 & 5 & 0 & 5 \\ 14 & 0 & 0 & 0 & 0 & 0 & 0 & 0 & 0 & 0 & 0 & 0 & 5 & 5 & 0 \end{array} ,$$

$$M_{final} = \begin{array}{c|cccccccccccccc} & 1 & 2 & 3 & 4 & 5 & 6 & 7 & 8 & 9 & 10 & 11 & 12 & 13 & 14 \\ \hline 1 & 0 & 1 & 1 & 0 & 0 & 0 & 0 & 0 & 0 & 0 & 0 & 0 & 0 & 0 \\ 2 & 1 & 0 & 1 & 0 & 0 & 0 & 0 & 0 & 0 & 0 & 0 & 0 & 0 & 0 \\ 3 & 1 & 1 & 0 & 0 & 0 & 0 & 0 & 0 & 0 & 0 & 0 & 0 & 0 & 0 \\ 4 & 0 & 0 & 0 & 0 & 1 & 1 & 0 & 0 & 0 & 0 & 0 & 0 & 0 & 0 \\ 5 & 0 & 0 & 0 & 1 & 1 & 0 & 0 & 0 & 0 & 0 & 0 & 0 & 0 & 0 \\ 6 & 0 & 0 & 0 & 1 & 1 & 0 & 0 & 0 & 0 & 0 & 0 & 0 & 0 & 0 \\ 7 & 0 & 0 & 0 & 0 & 0 & 0 & 0 & 0 & 0 & 0 & 0 & 0 & 0 & 0 \\ 8 & 0 & 0 & 0 & 0 & 0 & 0 & 0 & 0 & 0 & 0 & 0 & 0 & 0 & 0 \\ 9 & 0 & 0 & 0 & 0 & 0 & 0 & 0 & 0 & 0 & 1 & 1 & 0 & 0 & 0 \\ 10 & 0 & 0 & 0 & 0 & 0 & 0 & 0 & 0 & 1 & 0 & 1 & 0 & 0 & 0 \\ 11 & 0 & 0 & 0 & 0 & 0 & 0 & 0 & 0 & 0 & 1 & 1 & 0 & 0 & 0 \\ 12 & 0 & 0 & 0 & 0 & 0 & 0 & 0 & 0 & 0 & 0 & 0 & 0 & 1 & 1 \\ 13 & 0 & 0 & 0 & 0 & 0 & 0 & 0 & 0 & 0 & 0 & 0 & 1 & 0 & 1 \\ 14 & 0 & 0 & 0 & 0 & 0 & 0 & 0 & 0 & 0 & 0 & 0 & 1 & 1 & 0 \end{array}$$

Step-5: A community consists of all those elements in matrix whose value is equal to 1. Therefore, the edge betweenness algorithm applied on the graph (Figure 2.6) detected four communities as follows: C1 = {node 1, node 2, node 3}; C2 = {node 4, node 5, node 6}; C3 = {node 9, node 10, node 11}; and C4 = {node 12, node 13, node 14}, as shown encircled in Figure 2.6b.

2.2.2.7 Multilevel A multilevel algorithm [14, 15] detects the communities in a graph by computing the maximal matching. A maximal matching in a non-weighted graph G (V, E) is a subset of E. A maximal matching in a weighted graph G (V, E) occurs if the sum of two connected edges in E is equal to the joining edge. That is, node i and node j are combined, iff $A_{i,j} \equiv A_{i,k} + A_{k,j}$, where A is the adjacency matrix of the input graph. After the merger of node *i* and node *j*, they are considered to be a single node. Then communities are detected based on independent set nodes. An independent set of nodes of a graph G (V, E) is a subset of V such that no two nodes are connected by an edge. Algorithm 2.7 gives the pseudo-code of Multilevel.

ALGORITHM 2.7:   MULTILEVEL

Input: A graph $G = (V, E)$
Output: Communities, C

1. *Compute weighted Adjacency Matrix,* A_{ij}
2. *For* $i \in \{1 \ to \ n \} \forall \ V$
3. *For* $j \in \{1 \ to \ n \} \forall \ V$
4. *if* $A[i][j] == A[i][k] + A[k][j]$
5. $V_t = Merge(i,j)$
6. *end For*
7. *end For*
8. $C = independentsets \ \{V\}$

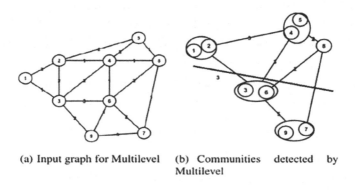

(a) Input graph for Multilevel   (b) Communities detected by Multilevel

Figure 2.7   a) and (b): Example for Multilevel

If there is an edge from node i to node j with weight w, then $A_{i,j} = w$, else $A_{i,j} = 0$. If $A_{i,j} == A_{i,k} + A_{k,j}$, then merge maximal matching vertices (i,j). The merger of nodes is updated in the adjacency matrix. Then partition the graph based on independent set nodes. One example input for Algorithm 2.7 is shown in Figure 2.7.

Step-1: The adjacency matrix of Figure 2.7a is:

$$A = \begin{bmatrix} 0 & 3 & 1 & 0 & 0 & 0 & 0 & 0 & 0 \\ 3 & 0 & 2 & 1 & 2 & 0 & 0 & 0 & 0 \\ 1 & 2 & 0 & 2 & 0 & 3 & 0 & 0 & 2 \\ 0 & 1 & 2 & 0 & 3 & 1 & 0 & 1 & 0 \\ 0 & 2 & 0 & 3 & 0 & 0 & 0 & 1 & 0 \\ 0 & 0 & 3 & 1 & 0 & 0 & 2 & 2 & 1 \\ 0 & 0 & 0 & 0 & 0 & 2 & 0 & 1 & 3 \\ 0 & 0 & 0 & 1 & 1 & 2 & 1 & 0 & 0 \\ 0 & 0 & 2 & 0 & 0 & 1 & 3 & 0 & 0 \end{bmatrix},$$

$$A_1 = \begin{bmatrix} & (1,2) & (3,6) & (4,5) & (7,9) & 8 \\ (1,2) & 0 & 3 & 3 & 0 & 0 \\ (3,6) & 3 & 0 & 3 & 5 & 0 \\ (4,5) & 3 & 3 & 0 & 0 & 2 \\ (7,9) & 0 & 3 & 0 & 0 & 1 \\ 8 & 0 & 2 & 2 & 1 & 0 \end{bmatrix}$$

Step-2: Find maximal match nodes in matrix A. Since, $A_{1,2} = = A_{1,3} + A_{3,2} = 3$ implies node 1 and node 3 are maximal match nodes, hence these nodes (node 1 and node 2) are merged. For similar reasons, node 4 and node 5, node 3 and node 6, and node 7 and node 9 are also merged.

Step-3: Update adjacency matrix, A_1.

Step-4: To form the communities, we need to find independent set nodes in A_1. Since Node (1,2) and Node (7,9) do not share the same edge, Node (1,2) and Node (7,9) will be in two different communities. Similarly, this process continues to place independent nodes in different communities. Finally, Multilevel algorithm applied on the graph (Figure 2.7) detected two communities, C1 = {node 1,node 2,node 4,node 5,node 8} and C2 = {node 3, node 6, node 7, node 9}. The detected communities are shown encircled in Figure 2.7b.

2.2.2.8 Fast Greedy   The Fast Greedy algorithm [20] is a hierarchical bottom-up approach that works on the principle of modularity. This approach uses the greedy optimization strategy to optimize the modularity values of the detected communities. Initially, this algorithm considers each node of the input graph as a single community (singleton set). In each iteration, it considers a node randomly along with its connected neighboring nodes. Then the modularity of the randomly picked node is computed in association with each of its neighboring nodes. The association that gives maximum modularity results in the merging of the respective nodes to form a community. This procedure is repeated until there is no further improvement in the modularity of that community.

The working of Fast Greedy on a graph shown in Figure 2.8a is explained below:

Step 1: In the first iteration, each node has to be considered as one community. Supposing node-1 is picked randomly, hence, its corresponding neighboring nodes $\{2,3,5\}$ are also taken into consideration. Now modularity between node 1 and each of its corresponding neighboring nodes is calculated. It decides whether node 1 is part of any one of its neighboring nodes, provided it results in maximum modularity with that node. Modularity is defined as $1-\dfrac{k_i k_j}{|E|}$, where k_i, k_j are the degrees of node i and node j, respectively. $|E|$ is the total number of edges in the graph. The modularity of node 1 with neighboring nodes $\{2,3,5\}$ is calculated as follows:

$$(1,2)=1-\frac{3*5}{16}=\frac{1}{16};(1,3)=1-\frac{3*4}{16}=\frac{4}{16};(1,5)=1-\frac{3*3}{16}=\frac{7}{16};$$

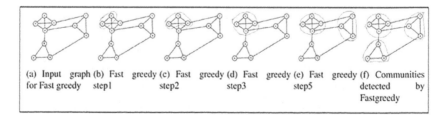

(a) Input graph for Fast greedy (b) Fast greedy step1 (c) Fast greedy step2 (d) Fast greedy step3 (e) Fast greedy step5 (f) Communities detected by Fastgreedy

Figure 2.8   (a)-(f): Example for Fast Greedy

So, the calculation above shows that node 1 and node 5 result in maximum modularity. Hence, nodes (1, 5) are merged to form one community, as shown in Figure 2.8b.

Step 2: Now (1,5) forms one community node and {2,3} are its neighboring nodes. So modularity of (1,5) with nodes {2,3} is calculated as follows:

$$(1,5,2) = 2 - \frac{4*5}{16} = \frac{12}{16}(1,5,3) = 2 - \frac{4*4}{16} = \frac{16}{16}$$

The calculations above show that community node (1,5) merged with node-3 results in maximum modularity. Hence, these three nodes (1,5,3) now form one community as shown in Figure 2.8c.

Step 3: Similarly the modularity calculation is carried out for community node (1,5,3) and its neighboring nodes {2,7}. The computations would show that community node (1,5,3), if merged with node-2, gives maximum modularity, hence nodes (1,5,3,2) form one community, as shown in Figure 2.8d.

Step 4: The modularity of community node (1,5,3,2) with each of its neighboring nodes {4,7,8} results in negative values as given below:

$$(1,5,3,2,4) = 1 - \frac{9*3}{16} = -ve; (1,5,3,2,7) = 1 - \frac{9*3}{16}$$

$$= -ve \text{ value}; (1,5,3,2,8) = 1 - \frac{9*1}{16} = -ve \text{ value}$$

Hence, no merger happens.

Step 5: Now Fast Greedy considers any random node from {4,7,8} and repeats the above-mentioned steps. Supposing node-4 is considered with its neighboring nodes {7,6,(1,5,3,2)}. This results in the merger of nodes (4,7) to form another community. Further, it leads to the merger of community node (4,7) with node-6, as shown in Figure 2.8e.

Steps 6–8: Continuing with such modularity computations, nodes (8,9,10) merge to form a community, as shown in Figure 2.8f. Therefore, the Fast Greedy algorithm applied on the graph (Figure 2.8a) detected three communities, C1 =

{node 1,node 2,node 3,node 5}, C2 = {node 4, node 6, node 7} and C3 = {node 8, node 9, node 10}. The detected communities are shown encircled in Figure 2.8f.

2.2.2.9 Louvain Louvain [13] is similar to the Fast Greedy algorithm. It has one additional step, called phase 2 (which represents the end of phase 1). In phase 2 it rebuilds the graph among the detected communities (considered as single nodes along with their connecting edges) for any further possibility of partitioning. If modularity between any of these communities is positive, then Steps 4 to 11 (of Algorithm 2.8) are repeated. In the case of negative it stops. This is shown in Figure 2.9.

ALGORITHM 2.8:   FAST GREEDY

Input: *A graph G = (V, E)*
Output: *Communities, C*

  1. *Compute Adjacency Matrix, A_{ij}*
  2. *Initialize $Q^{(0)}=0$ at i=0, // i- first iteration*
  3. *C_i = {i} ∈ {1 to n } ∀ V // Put each node of G in its community*
  4. *For i ∈ {1 to n } ∀ V*
  5. *N_i = {i} ∈ {1 to n } ∀ Neighboring$_n$odes(i) // place neighboring nodes of i in N*
  6. *For j ∈ {1 to m } ∀ N*
  7. *k_i = deg (node – i)*
  8. *k_j = deg (node – j)*
  9. *$Q^{(t)} = 1 - \dfrac{k_i * k_j}{|E|}$ // modularity*
  10. *if($Q^{(t)} > Q^{(t-1)}$)// if the new modularity is higher than the initial*
  11. *C_i = i ∪ j // Merge node i and node j as a one community*
  12. *End for*
  13. *End for*
  14. *C = C_i*

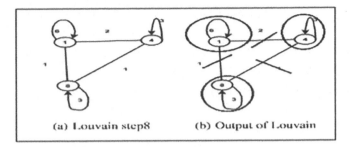

(a) Louvain step8   (b) Output of Louvain

Figure 2.9   (a) and (b): Phase 2 of Louvain

Step 9: Again modularity calculation has to be followed by considering the random node and its neighboring nodes. So, consider node-1 and its neighboring nodes $\{4, 8\}$. The modularity of node 1 with nodes $\{4, 8\}$ is as follows:$(1,4) = 2 - \dfrac{9*5}{16} = -ve$ value;$(1,8) = 1 - \dfrac{5*7}{16} = -ve$ value Since the modularity score is -ve, it means that the detected communities cannot be further divided into more communities. Therefore, the Louvain algorithm applied on the graph (Figure 2.9a) detected three communities, C1 = {node 1, node 2, node 3, node 5}, C2 = {node 4, node 6, node 7} and C3 = {node 8, node 9, node 10}. The detected communities are shown encircled in Figure 2.9b.

2.2.3 Gaps in Existing Community Detection Algorithms

This section highlights the gaps in the existing comparative study on community detection algorithms. Our comparative analysis of the existing community detection algorithms includes both disjoint and overlapping algorithms. Based on their topological, community, and structural properties, we identified high-performance algorithms. Existing surveys on community detection algorithms have mostly considered synthetic graphs. But we reviewed the algorithms on large-scale real-world datasets.

Danon et al. [20] performed a comparative study on different community detection algorithms based on their performances on the GN benchmark. But it has the following limitations: all nodes of the dataset are assumed to have some degree, communities are separated in the same way, and the network is unrealistically small.

Orman and Labatut [6] have done a comparative study on five community detection algorithms: *Louvain, Fast Greedy, Walktrap, Infomap,* and *Label Propagation.* The performance of these algorithms was compared based on the LFR benchmark and accuracy was computed on a smaller dataset (graphical representation contains 7500 to 250000 nodes) [33]. But they did not measure modularity and clustering coefficient for the time of execution. Whereas, Peel [23] compared two algorithms, Infomap and Label Propagation, on both weighted and unweighted graphs, comprising of only 100 nodes. The two algorithms are compared based only on the observed network parameters without having any prior knowledge of the community structure. So, such comparison lacks accuracy and correctness because of the smaller network size and the limited number of parameters.

Recently, Hric et al. [34] compared the accuracy of ten algorithms on real-world graphs (comprising of up to *5189809* nodes) based on LFR benchmark parameters. The studied algorithms are *Louvain, Infomap, CPM, Conclude, Label Propagation, Demon, Link communities,* and *Ganxis, InfomapSingle* (extension of Infomap), and *GreedyCliqueExp* (extension of CPM). These algorithms are compared, based on the non-topological features of the nodes. But this comparison is not detailed enough to suggest a suitable algorithm for a particular dataset based on its topological properties.

Some other research works on community detection [23] are employed on small artificial and/or real-world benchmark datasets, such as *Zachary's karate club network* and *Enron-email data.* However, as of now, there is no detailed study to clarify whether there is any correlation between the performance of any algorithm and the size of a network. The algorithms that have been proposed so far have not been evaluated either quantitatively or qualitatively on real-time graphs. Whereas, we systematically establish the dependency on network size based on the effective computation time of the algorithms.

2.3 Experimental Study

This section shows the experimental setup of our implementation, the results, i.e. the communities detected by the algorithms, and their comparison in terms of the computed parameters such as *computation time (CT), clustering coefficient (CC),* and *modularity(M).* These evaluation parameters quantify the association among vertices of the community. Due to the massive size of our dataset, clear visualization of all the detected communities is not feasible. The application of all the *community detection* algorithms and their behavioral study has been carried out on the graphical model of the dataset.

2.3.1 Framework

All the experiments are conducted on a computer server which is equipped with 4 Quad-Core AMD Opteron CPU, 32GB memory, and 12 T disk storage. The mint OS 6.4 is installed as the operating system. Other software environments include Python 3.6 and Pycharm 4.5.3.

The SN taken for this experimental study is the DBLP dataset.[1] It comprises of publication records (from 1936 to 2018) of different researchers published in various journals and conferences. This dataset is graphically modeled, wherein nodes correspond to the authors of a research paper and the edges correspond to the co-author relationships. Thus, a graphical representation is created for each year of DBLP data using the *networkx* library. The nodes of this graphical representation exhibit a constituent relation through the interactions among themselves [35]. Such an interconnected set of nodes having active interactions are characterized as communities [33]. The results shown in this chapter are the outcomes of our implementation on 18 years of DBLP data (i.e. data from the years 2000 to 2018). This is because, before the year 2000, the available data are sparse and contribute little to the formation of communities; hence it is not relevant for the analysis of community detection algorithms. In addition, the volume of data between 2000 and 2018 is quite high and hence suitable for the demonstration of algorithms' capacity in handling large, growing, datasets. The steps followed are:

1. Collect data from the *DBLP* dataset, which is in the <xml> format.
2. Process the data to store yearly information in pickle format (a serialized way of storing the extracted data in Python programming language).
3. Construct yearly graphs (considering data of each year), where authors form the nodes and co-author relationships, if any, form the edges.

4. Implement a community detection algorithm on all yearly graphs.
5. Compare the results based on LFR benchmark metrics.

Our proposed framework in the form of the above five steps is shown pictorially in Figure 2.10.

2.3.2 Result Analysis

This section gives an analysis of the results (detected communities) obtained for each community detection algorithm and answers the three RQs. For each algorithm, evaluation metrics are computed to look into the quality of communities detected by different methods. The computed evaluation metrics for each algorithm are shown in Figure 2.11, where the X-axis represents the corresponding year and the Y-axis represents the computed metrics. The CD algorithms are ranked based on these results and are shown later in Table 2.2. To overcome the effect of randomness, we implemented all these algorithms and executed them 50 times. The average of all the iteration results are considered for analysis and to wave out any form of bias. The general observation is that with the yearly increase in network size there is an increment in the *number*, *Clustering Coefficient*, and *Modularity* of communities. Now, with the results in hand let's look into the answers to our initial research questions.

2.3.2.1 RQ1. Are the Existing CD Algorithms Suitable for Evolving SNs? Here, we analyze the performance of the algorithms based on how their parameters (clustering coefficient, modularity, and computation time) changed each year. Our observations in this regard for each of the algorithms are as follows.

1. **Louvain** algorithm performance is represented with a blue color line in Figure 2.11. It performs best on large datasets compared to the remaining algorithms. It detected 31136 communities in the year 2017 in only 59.49 seconds with CC = 0.8352 and modularity = 0.8851. It can be seen from Table 2.1 that this is the fastest community algorithm when compared to remaining algorithms on large-scale networks. And it also gives stabilized CC and modularity values for 270900

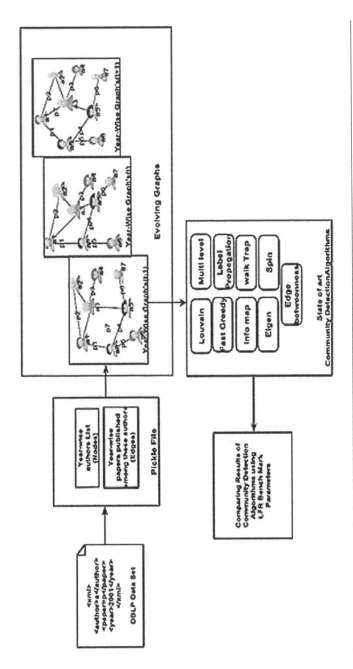

Figure 2.10   Framework for Analysis of Community Detection Algorithms

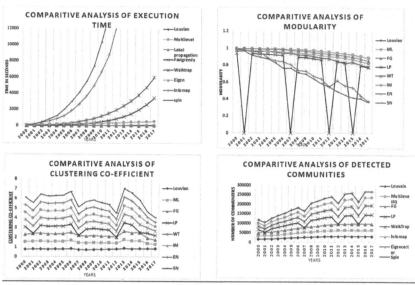

Figure 2.11   Comparative Analysis of Nine Community Detection Algorithms

number of nodes and 773264 number of the graph of 2017 DBLP SN. From the perspective of running time, it detects 31136 communities in 59.49 seconds. Our interpretation, from Figure 2.11, is that with an increase in the number of nodes and edges, the time of execution also increases. At the same time, it maintains consistent CC and Modularity. This is because of better modularity values at higher performance values of the Louvain Algorithm. Therefore, we conclude that Louvain is suitable for large scale datasets and is also scalable.

2. **Multilevel partitioning** also detects 31336 communities in 31.35377265 seconds with good modularity of 0.8860, but it has a lower CC value of 0.4543. This algorithm performs like the Louvain algorithm, but with CC compromised. It also detects communities in reasonable computation time on large datasets and gives stable modularity values for graphs comprising of 270900 nodes to 773264 edges. The performance is represented with a red color line in Figure 2.11. It is inferred that with the increase in the number of nodes and edges, there is an increase in the time of execution. It is also observed that while modularity remained consistent, but for some years CC measure was inconsistent. Multilevel clustering is suitable for

all those applications (such as *Google Maps*, *Topic Maps*, etc.) wherein clustering co-efficient is not vital.

3. **Label propagation**: The performance is represented with a violet color line in Figure 2.11. It is observed that this algorithm detected communities until the size of the graph reached 167492 nodes and 357813 edges (data of the year 2010). Its modularity values decreased thereafter for the remaining years. This indicates that it is not able to form the communities accurately when the size of the dataset grows beyond a limit. However, this approach gives the most satisfying results with good CC and modularity up to the year 2009 (smaller data) at the cost of a little higher time of 347.33 seconds compared to above mentioned two algorithms.

4. **Fastgreedy method**: It detected the communities with a good modularity score of 0.8499 for the graph (of the year 2017) comprising of 270900 nodes and 773264 edges. These results are shown in Figure 2.11 with a green color line. This algorithm is stable for large networks if only modularity is considered and performs well for small and medium-sized networks if computation time and modularity are considered. For large networks, the performance is not comparable with Louvain and multi-level communities. It is pretty fast and merges nodes in linear time for small and medium-sized graphs. Even smaller communities tend to merge even if they are well-grouped within large networks. Therefore, this algorithm fails to correctly detect communities in large datasets. Another problem here is that it cannot detect communities for non-integer values of modularity. This limitation was observed from the results of the dataset of the year 2001, 2007, 2012, and 2015.

5. **Walktrap algorithm**: The results in Figure 2.11 with a light blue color line show that this algorithm takes more time to detect the communities on a large set of graphs. The modularity score was good for the data of the years 2000 to 2010, with CC being mostly inconsistent. The CC and modularity values started dropping for those graphs with more than 196374 nodes and 438687 edges (i.e., for the data of the year 2011 onwards). Hence, we observed that it is suitable for a smaller

network whose graphical representation will not contain more than 167492 nodes.

6. **Infomap**: The performance of *Infomap* is shown in Figure 2.11 with an orange color line. Infomap algorithm performs well in terms of community quality, but the time complexity increases (because of the overhead of encoding the information) with the increase in network size. It works similarly to the Walktrap algorithm for the graph (of the year 2012) with 196374 nodes and 438687 edges. After that, as the size of the graph increased, the encoding of the data took more time. As a result, the time to detect communities also increased. So, this method is suitable for graphs with up to 182373 nodes and 402861 edges.

7. **Leading eigenvector**: It uses a top-down approach that seeks to maximize the modularity to combine a similar set of groups. It decomposes the modularity matrix to detect a set of communities. But the time taken to detect the communities is high, i.e. it detects communities efficiently but at the cost of time. It can be inferred from Figure 2.11 with a dark blue color line. It can also be observed that this approach is slower than the Fast Greedy method and not stable for graphs with not more than 71165 nodes and 120969 edges (i.e. it is not suitable for data of the year 2003 and thereafter).

8. **Spinglass**: This method working process depends on the Hamiltonian Matrix . For small datasets, it is detecting the correct number of communities. We verified it for years 1960 to 1980 after that it is detecting only 25 communities. Because for large data-sets, in 25 steps, that Hamiltonian Matrix reaches null. So, in our implementation, it detected only 25 communities for all the years from 2000 to 2017. Moreover, this method is not fast and deterministic because of the simulation of spins. This analysis is shown in Figure 2.11 with the maroon color line.

9. **Edge Betweenness**: This method yielded results only for a small set of graphs with 700 nodes and 3500 edges. We tested this method on the DBLP dataset for both large and small graphs. In a graph (of the year 1970) with 1692 nodes and 748 edges, this method detected only 1 community with CC =

0.850125. Similarly, we tested this method on a graph (of the year 1990) with 6041 nodes and 15851 edges, it took 2567.52 minutes to detect only 5 communities with CC = 0.8035. Hence, our observation is that this method is not suitable for the detection of communities in large datasets.

2.3.2.2 RQ 2. Are the LFR Benchmark Parameters Suitable for Comparing CD Algorithms? The computational values of the metrics of CD algorithms are listed in Table 2.1. Table 2.1 shows each algorithm CC (Clustering Coefficient), M (Modularity), and CT (Computation Time). The *modularity (M)* is computed for all the nine algorithms for the data belonging to the years 2000 to 2017. The modularity of *Louvain (M = 0.88)*, *Multilevel (M = 0.88)*, *Fastgreedy (M = 0.84)*, *Infomap (M = 0.82)*, and *Label Propagation (M = 0.77)* performed well for large datasets, for graphs which has up to 270900 nodes and 773264 edges. The boxplot of this comparative analysis is given in Figure 2.12. Based on the above observations, we infer instead of modularity that both Louvain and Multilevel perform well for the DBLP dataset.

The *clustering coefficient (CC)* of every algorithm for data belonging to the years 2000 to 2017. The Louvain algorithm gives a consistent performance when the size of the graphs increased for every year. The performance of remaining algorithms depended on termination conditions, i.e. the number of iterations required to detect the communities. All algorithms (except Louvain) performed well until the graph (of the year 2015), with up to 234491 nodes and 658158 edges. Eigenvector and Spinglass have inconsistent CC values for large datasets. Figure 2.12 shows that CC values of Louvain are distributed uniformly for small to large datasets. Whereas remaining algorithms left-skewed the distribution values, i.e. CC values are distributed uniformly for small datasets and failed to detect accurate communities for large datasets.

Similarly, the Computation Time (in seconds) of each algorithm for detecting the communities in the data of the years 2000 to 2017. Louvain (Time = 59.49), Multilevel (Time = 31.835), Fast Greedy (Time = 18.05), and Infomap (Time = 533.54) detected communities in less time for large datasets (for graphs with up to 270900 nodes and

TABLE 2.1 Comparative Analysis of Community Detection Algorithms Based on Modularity (M), Clustering Coefficient (CC) and Computation Time (CT)

YEAR	NODES	EDGES	LOUVAIN			MULTILEVEL			FASTGREEDY			LABELPROP.		
			CC	M	CT	CC	M	CT	CC	M	CT	CC	M	CT
2000	52323	73629	0.76	1	0.60	0.76	1	0.42	0.76	0.99	6.14	0.94	0.76	0.472
2001	55791	78912	0.79	1	1.51	0.79	1	0.82	0.79	0.99	12.71	nil	nil	nil
2002	60448	91441	0.8	0.99	2.37	0.8	0.99	1.29	0.8	0.99	21.19	0.94	0.8	0.6589
2003	71165	120969	0.78	0.99	3.64	0.78	0.99	1.89	0.78	0.99	32.28	0.93	0.78	1.2412
2004	79760	132031	0.78	0.99	4.92	0.78	0.99	2.53	0.78	0.99	46.80	0.93	0.78	2.1819
2005	96321	176028	0.79	0.99	6.78	0.79	0.99	3.36	0.79	0.99	71.58	0.92	0.79	3.7656
2006	108542	204009	0.84	0.99	8.58	0.84	0.99	4.37	0.84	0.98	119.46	0.92	0.84	5.626
2007	122922	228674	0.73	0.99	10.87	0.73	0.99	5.48	0.73	0.98	176.08	nil	nil	nil
2008	135853	265521	0.71	0.98	12.99	0.71	0.98	6.83	0.71	0.97	250.52	0.9	0.71	1.9038
2009	151871	312321	0.73	0.98	15.76	0.73	0.98	8.30	0.73	0.97	347.82	0.89	0.73	4.7948
2010	167492	357813	0.75	0.97	19.59	0.69	0.97	10.18	0.69	0.95	476.48	0.88	0.69	8.116
2011	182373	402861	0.77	0.96	23.34	0.66	0.96	12.24	0.66	0.94	639.61	0.86	0.66	12.745
2012	196374	438687	0.79	0.95	27.08	0.61	0.96	14.42	0.61	0.93	840.94	nil	nil	nil
2013	208958	526187	0.88	0.94	32.19	0.88	0.94	17.07	0.88	0.92	1106.94	0.85	0.88	4.7819
2014	224738	599177	0.82	0.93	37.53	0.82	0.93	20.00	0.82	0.91	1476.79	0.83	0.82	12.013
2015	234491	658158	0.83	0.92	44.37	0.83	0.92	23.39	0.83	0.9	1968.47	nil	nil	nil
2016	252771	690372	0.83	0.9	52.02	0.54	0.9	26.98	0.54	0.88	2592.85	0.8	0.54	8.2598
2017	270900	773264	0.84	0.89	59.49	0.45	0.89	31.35	0.45	0.85	3433.06	0.77	0.45	18.046

(*Continued*)

TABLE 2.1 (CONTINUED) Comparative Analysis of Community Detection Algorithms Based on Modularity (M), Clustering Coefficient (CC) and Computation Time (CT)

YEAR	NODES	EDGES	LOUVAIN			MULTILEVEL			FASTGREEDY			LABELPROP.		
			CC	M	CT	CC	M	CT	CC	M	CT	CC	M	CT
2000	52323	73629	0.76	0.98	19.60	0.76	0.98	89.93	0.76	0.99	82.92	0.76	0.96	89.56
2001	55791	78912	0.79	0.98	41.97	0.79	0.98	202.29	0.79	0.98	95.46	0.79	0.97	197.05
2002	60448	91441	0.8	0.97	69.64	0.8	0.97	403.49	0.8	0.96	111.69	0.8	0.92	451.33
2003	71165	120969	0.78	0.96	108.18	0.78	0.97	577.59	0.78	0.96	151.37	0.78	0.9	816.76
2004	79760	132031	0.78	0.96	159.73	0.78	0.97	854.35	0.78	0.89	156.98	0.78	0.88	1192.99
2005	96321	176028	0.79	0.95	235.86	0.79	0.96	1194.25	0.79	0.85	170.79	0.79	0.83	1959.84
2006	108542	204009	0.84	0.94	338.59	0.84	0.96	1826.54	0.84	0.84	203.78	0.84	0.77	2923.51
2007	122922	228674	0.73	0.94	477.05	0.73	0.95	2624.03	0.73	0.81	228.47	0.73	0.77	4051.40
2008	135853	265521	0.71	0.92	652.74	0.71	0.94	3656.25	0.71	0.74	242.76	0.71	0.71	5427.85
2009	151871	312321	0.73	0.92	877.16	0.73	0.94	4900.65	0.73	0.73	263.35	0.73	0.69	7270.28
2010	167492	357813	0.69	0.9	1159.89	0.69	0.93	6608.66	0.69	0.67	297.51	0.69	0.64	10255.82
2011	182373	402861	0.66	0.88	1507.56	0.66	0.91	8687.46	0.66	0.64	323.34	0.66	0.59	13683.80
2012	196374	438687	0.61	0.87	1927.18	0.61	0.9	12101.61	0.61	0.54	335.93	0.61	0.54	17527.73
2013	208958	526187	0.88	0.84	2431.23	0.88	0.89	15684.69	0.88	0.62	408.10	0.88	0.49	21681.77
2014	224738	599177	0.82	0.82	3058.58	0.82	0.88	20069.47	0.82	0.56	446.10	0.82	0.45	26981.11
2015	234491	658158	0.83	0.83	3787.29	0.83	0.87	25850.46	0.83	0.53	474.56	0.83	0.42	31506.54
2016	252771	690372	0.54	0.8	4726.53	0.54	0.84	33712.20	0.54	0.44	507.23	0.54	0.41	37892.45
2017	270900	773264	0.45	0.78	5996.44	0.45	0.82	42083.32	0.45	0.37	533.541	0.45	0.37	44592.91

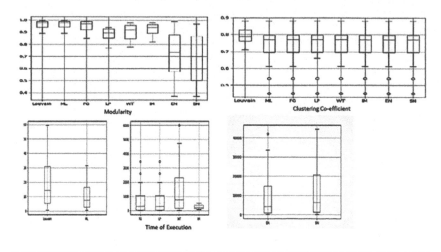

Figure 2.12   Comparative Analysis of Community Detection Algorithms

773264 edges). The execution time of the algorithms depended on the representation of graphs, criteria to detect the communities, and modularity calculation.

2.3.2.3 RQ 3. If So, Which Metrics are More Significant for Evolving SNs?   We further analyze the results in four aspects:

First, if computing time is not of concern (in the case of small networks), one should choose algorithms based on their *Modularity* and *CC.* In such a case, among all the algorithms, *Infomap, Label Propagation, Multilevel, Walktrap, Spinglass,* and *Edge Betweenness* algorithms are efficient to successfully uncover the structure of small networks.

Second, in case of network size is large or evolving, the results show that *Spinglass, Label propagation,* and *Edge Betweenness* algorithms exhibited lesser CC compared to *Louvain, Multilevel, Walktrap,* and *Infomap* algorithms. So, for large-scale datasets *Louvain, Multilevel, Walktrap,* and *Infomap* algorithms are suitable.

Third, in the case of large networks, it is suggested that algorithms should be chosen that can detect the communities in a reasonable time. From this point of view *Fast Greedy, Multilevel* and *Louvain* algorithms performed well.

Fourth, in the case of large networks, taking better *Modularity, CC*, and *Computation Time* into account, *Louvain* exhibited better performance compared to all other algorithms.

2.3.2.3.1 The rank of Algorithm   To answer existing CD algorithms suitable for different sizes of datasets, we come with an intuitive way for comparing measures based on their metric ranking. We summarized these metric rankings of each parameter of the algorithm into the ranking of each algorithm as *Rank_Algo(i)*.

$$Rank\_Algo(i) = \frac{1}{n}\sum_{j=1}^{j=n} Rank\_j(i).$$ Where i is the algorithm and j

is the rank of the parameter. Consider Louvain, to get the Rank of Louvain, substitute individual rank metrics of Louvain into this equation, where i=Louvain and n=4 (number of metrics):

$$Rank\_Algo(Louvain) = \frac{1}{4}\sum_{j=1}^{j=4} Rank\_j(i) = \frac{1}{4}Rank\_CD(Louvain)$$
$$+ Rank\_M(Louvain) + Rank\_CC(Louvain)$$
$$+ Rank\_T(Louvain) = \frac{1}{4}(1+1+1+1+2) = 1.$$

Similarly, we calculated the Rank of each algorithm and represented in Table 2.2 with the column name Rank_Algo. From Table 2.2, we observed that the top three algorithms (Louvain, Multilevel, and Fast Greedy) outperform for large datasets. The next three algorithms (Eigenvector, Infomap, and Walktrap) are exclusively for medium dataset and the last three algorithms (Label Propagation, Spinglass, and Edge Betweenness) are suitable for small datasets.

TABLE 2.2   Ranking of Community Detection Algorithms Based on LFR Benchmark Parameters

ALGORITHM	RANK_CD	RANK_M	RANK_CC	RANK_T	RANK_ALGO
Louvain	1	1	1	2	1
Multilevel	3	2	2	1	2
Fast Greedy	4	3	3	3	3
Eigenvector	2	7	6	6	5
Infomap	5	4	5	5	5
Walktrap	6	6	4	4	5
Label Propagation	7	5	8	8	7
Spinglass	8	8	7	7	8
Edge Betweenness	9	9	9	9	9

Hence, to answer RQ_3, the community detection algorithms are ranked (in the order of high to low) based on their modularity, clustering coefficient, computation time, as shown in Table 2.2. To justify this statement, we considered the boxplots of Modularity, CC, and Computation Time shown in Figure 2.12. This figure summarizes the statistical distribution of the Modularity, CC, and Computation Time values. By observing the median of these values, we can conclude that Louvain has the best values for CC and Modularity. In the same way, we calculated Modularity, CC, and Computation Time for the remaining algorithms. Based on their execution time, we categorized these algorithms into three sets. The first set represents algorithms with less execution time (less than 60.00 seconds), such as Louvain, Multilevel, and Fastgreedy. The second set represents algorithms with medium execution time (less than 6000.00 seconds), such as Label Propagation, Walktrap, and Infomap. And the third set represents more execution time algorithms (more than 1 hour) such as Eigenvector and Spinglass. From the boxplot (Figure 2.12), we observe that these three are right-skewed figures. It is obvious that algorithms take more time for large datasets. So, all these algorithms observed this rule and hence they are right-skewed. Therefore, from the analysis of the experimental results, we infer that Louvain is the most appropriate algorithm for the detection of communities in large and evolving datasets such as DBLP. We have also observed that the performance of these algorithms would vary for datasets with different features.

2.3.3 Threats to Validity

Based on our work, we have advanced some meaningful answers to the proposed three research questions. However, there are still some potential limitations and threats to the validity of our work.

1. Apart from the dataset used in our experimental analysis, there exist many other co-author datasets (such as arxiv, Pubmed, Thomson Reuters, etc.) that can be represented as a SN. Collectively, all these datasets would make a very voluminous dataset and experiments on such a combined dataset will certainly give different insights. But we considered only one dataset because of our limitations with computing power

and storage space. Another reason for going with the selected database is that it stores yearly data and exhibits co-author collaboration by maintaining power-law distribution between paper count and the number of authors for each year.

2. As the results, such as time of computation, depends upon the underlying system configurations and capabilities, in a high-performance computing facility the results would vary from the ones shown in this survey. But it will not affect the final experimental inference.

3. LFR benchmark is a generalized version of the earlier GN benchmark, having some extra features such as degree of distribution, modularity, CC, and community size. We felt the need of having more realistic properties. Since our implementation has been based on evolving temporal graphs, hence, consideration of graph-based properties, such as *average degree distribution*, *degree distribution exponent*, and *community distribution exponent* may impact the comparison of algorithms.

2.4 Conclusion and Future Work

In this survey, we compared the results of nine community detection algorithms using LFR benchmark metrics to answer our research questions. A suitably large dataset that was extracted from the DBLP database was used for the implementation of algorithms. This dataset was represented in the form of graphs to carry out the experimental analysis. We observed that Spinglass and Edge Betweenness performed well for smaller datasets. But these two did not scale well for large datasets. There are algorithms, such as *Infomap*, *Fast Greedy*, *Walktrap*, and *Multilevel*, which scaled well with large datasets but detected communities with low modularity. Algorithms like *Louvain*, *Label Propagation*, and *Eigenvector* are scalable for large datasets, and the detected communities (by these algorithms) had similar modularity. Among these three algorithms, *Louvain* is the fastest community detection algorithm followed by *Label Propagation* and *Eigenvector* algorithms. The answers to the research questions and the experimental results shown in this survey work will help the researchers in selecting an appropriate community detection algorithm.

The implementation results show that there is scope to improve the performance of existing algorithms. In the future, we aim to work on the parallelization of the existing algorithms to achieve faster computational results. We would also study the efficiency of meta-heuristic algorithms in the detection of communities in large evolving networks and mine the detected communities for events that cause these communities to evolve.

Note

1 https://dblp.uni-trier.de/

References

1. Jierui Xie, and Boleslaw K. Szymanski. 2011. "Community Detection Using a Neighborhood Strength Driven Label Propagation Algorithm." arXiv Preprint arXiv:1105.3264.
2. Seunghyeon Moon, Jae-Gil Lee, Minseo Kang, Minsoo Choy, and Jinwoo Lee. 2016. "Parallel Community Detection on Large Graphs with Mapreduce and Graphchi." *Elsevier Data & Knowledge Engineering*, pp. 17–31.
3. Jure Leskovec, Jon Kleinberg, and Christos Faloutsos. 2005. "*Graphs over Time: Densification Laws, Shrinking Diameters, and Possible Explanations.*" In *ACM Proceedings of the Eleventh SIGKDD International Conference on Knowledge Discovery in Data Mining*, pp. 177–187.
4. Mark EJ Newman, and Michelle Girvan. 2004. "Finding and Evaluating Community Structure in Networks." *Physical Review E* 69(2), pp. 1056–1079.
5. Martin Rosvall, and Carl T. Bergstrom. 2007. "An Information-Theoretic Framework for Resolving Community Structure in Complex Networks." *Proceedings of the National Academy of Sciences* 104(18), pp. 7327–7331.
6. Günce Keziban Orman, Vincent Labatut, and Hocine Cherifi. 2012. "Comparative Evaluation of Community Detection Algorithms: A Topological Approach." *Journal of Statistical Mechanics: Theory and Experiment* 2012(8), P08001.
7. Aaron Clauset, Mark E.J. Newman, and Cristopher Moore. 2004. "Finding Community Structure in Very Large Networks." *Physical Review E* 70(6), 066111.
8. Pascal Pons, and Matthieu Latapy. 2005. "*Computing Communities in Large Networks Using Random Walks.*" In *Springer International Symposium on Computer and Information Sciences*, pp. 284–293.
9. Natarajan Nagarajan, Prithviraj Sen, and Vineet Chaoji. 2013. "*Community Detection in Content-Sharing Social Networks.*" In *ACM International Conference on Advances in Social Networks Analysis and Mining*, pp. 82–89.

10. Lars Backstrom, Dan Huttenlocher, Jon Kleinberg, and Xiangyang Lan. 2006. *"Group Formation in Large Social Networks: Membership, Growth, and Evolution."* In *Proceedings of the 12th ACM SIGKDD International Conference on Knowledge Discovery and Data Mining*, pp. 44–54.

11. Chakrabarti Deepayan, Ravi Kumar, and Andrew Tomkins. 2006. *"Evolutionary Clustering."* In *Proceedings of the 12th ACM Sigkdd International Conference on Knowledge Discovery and Data Mining*, pp. 554–560.

12. Tanmoy Chakraborty, Suhansanu Kumar, Niloy Ganguly, Animesh Mukherjee, and Sanjukta Bhowmick. 2016. "GenPerm: A Unified Method for Detecting Non-Overlapping and Overlapping Communities." *IEEE Transactions on Knowledge and Data Engineering* 28(8), pp. 2101–2114.

13. Vincent D. Blondel, Jean-Loup Guillaume, Renaud Lambiotte, and Etienne Lefebvre. 2011. "The Louvain Method for Community Detection in Large Networks." *Journal of Statistical Mechanics: Theory and Experiment* 10, P10008.

14. Martin Rosvall, Daniel Axelsson, and Carl T. Bergstrom. 2009. "The Map Equation." *The European Physical Journal Special Topics* 178(1), pp. 13–23.

15. Hanlin Sun, Wei Jie, Christian Sauer, Sugang Ma, Gang Han, Zhongmin Wang, and Kui Xing. 2017. *"A Parallel Self-Organizing Community Detection Algorithm Based on Swarm Intelligence for Large Scale Complex Networks."* In *IEEE 41st Annual Computer Software and Applications Conference (Compsac)*, pp. 806–815.

16. Vincent D. Blondel, Jean-Loup Guillaume, Renaud Lambiotte, and Etienne Lefebvre. 2008. "Fast Unfolding of Communities in Large Networks." *Journal of Statistical Mechanics: Theory and Experiment* 2008(10), P10008.

17. Usha Nandini Raghavan, Réka Albert, and Soundar Kumara. 2007. "Near Linear Time Algorithm to Detect Community Structures in Large-Scale Networks." *Physical Review E* 76(3), 036106.

18. Rahil Sharma, and Suely Oliveira. 2017. "Community Detection Algorithm for Big Social Networks Using Hybrid Architecture." *Big Data Research* 10, pp. 44–52.

19. Vinay Singh, Anurag Singh, Divya Jain, Vimal Kumar, and Pratima Verma. 2018. "Patterns Affecting Structural Properties of Social Networking Site Twitter." *International Journal of Business Information Systems* 27(4), pp. 411–432.

20. Danon Leon, Diaz-Guilera Albert, Duch Jordi, Arenas Alex. 2005. "Comparing Community Structure Identification". *Journal of Statistical Mechanics: Theory and Experiment*, P09008.

21. Jörg Reichardt, and Stefan Bornholdt. 2006. "Statistical Mechanics of Community Detection." *Physical Review E* 74(1), 016110.

22. Ery Arias-Castro, Gilad Lerman, and Teng Zhang. 2017. "Spectral Clustering Based on Local PCA." *The Journal of Machine Learning Research* 18(1), pp. 253–309.

23. Leto Peel. 2010. *"Estimating Network Parameters for Selecting Community Detection Algorithms."* In *IEEE 13th Conference on Information Fusion (Fusion)*, pp. 1–8.

24. Matthew J. Rattigan, Marc Maier, and David Jensen. 2007. *"Graph Clustering with Network Structure Indices."* In *ACM Proceedings of the 24th International Conference on Machine Learning*, pp. 783–790.
25. Xiang Li, Yong Tan, Ziyang Zhang, and Qiuli Tong. 2016. *"Community Detection in Large Social Networks Based on Relationship Density."* In *IEEE 40th Annual Computer Software and Applications Conference (Compsac)*, pp. 524–533.
26. Mark EJ Newman. 2006. "Finding Community Structure in Networks Using the Eigenvectors of Matrices." *Physical Review E* 74(3), 036104.
27. Pradeep Kumar, Kishore Indukuri Varma, Ashish Sureka, and others. 2011. "Fuzzy Based Clustering Algorithm for Privacy-Preserving Data Mining." *International Journal of Business Information Systems* 7(1), pp. 27–45.
28. Andrea Lancichinetti, Santo Fortunato, and Filippo Radicchi. 2008. "Benchmark Graphs for Testing Community Detection Algorithms." *Physical Review E* 78(4), 046110.
29. Rajmonda Sulo Caceres, and Tanya Berger-Wolf. 2013. "Temporal Scale of Dynamic Networks." In *Temporal Networks* (Springer, Berlin, Heidelberg), pp. 65–94.
30. Gergely Palla, Imre Derényi, Illés Farkas, and Tamás Vicsek. 2005. "Uncovering the Overlapping Community Structure of Complex Networks in Nature and Society." *Nature* 435(7043), N814.
31. Randolf Rotta, and Andreas Noack. 2011. "Multilevel Local Search Algorithms for Modularity Clustering." *ACM Journal of Experimental Algorithmics (JEA)*, 16, pp. 2–13.
32. Paul W. Holland, and Samuel Leinhardt. 1971. "Transitivity in Structural Models of Small Groups." *Comparative Group Studies* 2(2), pp. 107–124.
33. Andrea Lancichinetti, Mikko Kivelä, Jari Saramäki, and Santo Fortunato. 2010. "Characterizing the Community Structure of Complex Networks." *PloS One* 5(8), e11976.
34. Darko Hric, Richard K. Darst, and Santo Fortunato. 2014. "Community Detection in Networks: Structural Communities Versus Ground Truth." *Physical Review E* 90(6), 062805.
35. Michelle Girvan, and Mark E.J. Newman. 2002. "Community Structure in Social and Biological Networks." *Proceedings of the National Academy of Sciences* 99(12), pp. 7821–7826.

3

WHY THIS RECOMMENDATION?

Explainable Product Recommendations with Ontological Knowledge Reasoning

VIDYA KAMMA

KL University, Guntur, Andhra Pradesh

D. TEJA SANTOSH

Sreenidhi Institute of Science and Technology,
Ghatkesar, Telangana

SRIDEVI GUTTA

Gudlavalleru Engineering College
Gudlavalleru, Vijayawada, Andhra Pradesh

Contents

3.1 Introduction

In previous years, e-commerce product recommendations were the driving force in enabling users to make wise purchase decisions [1]. In today's world of smart decision-making, however, the users are in no position to accept only the displayed recommendations [2]. This is because users have expressed that these recommendations merely lack [3] the convincing explanation of why this product emerged as top of the list of recommended products. This has led to the concept of using the recommender system to explain [4] why this recommendation with the term "Explainable Recommendation" [5]. The explanations of product recommendations are the reasons [6] to support the "why-this" aspect of intelligent predictions made by the recommender system. These reasons increase both the degree of transparency and users' trust in these systems.

In order to generate explanations by the recommender system the statistical knowledge obtained from product reviews, namely product aspects, corresponding sentiments and statistical recommendations, are annotated in the engineered ontology as individuals to the developed classes. The semantic web rules built on top of the ontology helps the recommender system to find the knowledge paths as associated explanations to the recommendations. These knowledge paths are derived from the users about the products in the ontology data model. This way of knowledge-based embedding for explainable product recommendations provides the users with the awareness of why they are seeing such recommendations. This also improves the user satisfaction for recommender systems.

3.2 Related Works

In recent years, the explanations of recommender systems have developed into an active research area [7]. In particular, research about the explanation of which products form a part of the base cases in Case Based Reasoning (CBR) has gained considerable interest in recent times [8]. In recommending the products, product reviews are a key indicator. The works [9] on quantified sentiments on aspects from reviews in the formation of CBR cases for determining similarity based on recommendations forms the basis for explaining the recommendations. Firstly, works on deep learning architectures are surveyed for the foremost important task of aspect extraction. Next, there is a survey of the literature on opinions and sentiments of the extracted aspects.

In addition, there is an exploration of the literature on generating experiential cases by using CBR for recommendations. Finally, the chapter contains an examination of the literature on explaining the determined product recommendations through the use of various techniques.

Soujanya et al. used [10] a seven-layer-deep Convolutional Neural Network (CNN) in which they their developed linguistic patterns are used as an input to extract the words in the reviews as aspects. Ye et al. introduced [11] convolutions of operation over sentence dependency parse tree in order to mine the aspects from the reviews. Luo et al. offered [12] a deep neural network by integrating bi-directional Long Short-Term Memory (BiLSTM), Conditional Random Field (CRF) and word embeddings into the network for extracting the aspects. In another contribution, Xu et al. utilized [13] a double embedding method in CNN for aspect extraction. Aminu and Salim used [14] word embeddings and POS tags as inputs to CNN for aspect extraction. These researchers used these as two input channels. To the best of their knowledge, the authors of this current chapter specify that domain-adapted word embeddings and the lexicons with the word embeddings in addition to POS tags improves the performance of aspect extraction using the modified CNN model.

Bolanle and Olumide extracted [15] the opinions by using a relation-based approach. Santosh et al. mined [16] the opinions by using the positive and negative seed sets and comparing the adjectives, adverbs and verbs with these seed sets in order to identify potential matches. The matched words are deemed as opinions. Recently, KC Ravi Kumar et al. modified [17] the distance metric for determining the Semantic Orientation (SO) of the opinions. They used Sentiwordnet scores to quantify the extracted opinions, used them in the modified distance metric and calculated opinion orientations. The research on extracting aspects and automatic classification of the opinions with the help of deep learning models had drawn a keen interest during the year 2016. Lakkaraju et al. proposed [18] a hierarchical deep learning architecture to extract both aspects and classified the opinions through the use of word vectors and matrices. Gu et al. proposed [19] cascaded CNN to tackle aspect extraction and opinion orientation. Recently, Luo et al. proposed [20] a new Recurrent Neural Network (RNN) in order to extract aspects and their polarities in terms of the co-extraction task. The determination of the sentiment of the aspect with their counted opinion orientations are performed by using the Sentiment measure.

The case-based product recommendation which utilized the extracted aspects and sentiments gained its importance in the year 2013. In that year, Dong et al. described [1] a new approach to case-based recommendation that takes advantage of aspects and sentiments extracted from reviews and automatically generates both query case and base cases. In this process, the recommendation is the outcome based on the common aspects sentiments-based similarity between query case and base cases. Recently, Teja et al. utilized [21] CBR in order to arrive at product recommendations.

Research into explanations of the recommendations has gained popularity in recent years. Catherine et al. illustrated [22] the generation of explanations by using the external knowledge as knowledge graphs. Ai et al. proposed [23] to adopt knowledge graph embeddings for explainable recommendation. The authors constructed a user-item knowledge graph, embedding various user, item, and entity relationships. Abdollahi and Nasraoui built [24] the movie recommender system, which explains the recommendations by using the Constrained Matrix factorization technique. However, they have built their explainable recommender system by utilizing the movie ratings given by the users. The research work discussed above in [22] has used a first-order probabilistic logic system with a Page Rank approach to produce the ranked list of entities as explanations. This system finds metapaths as explanation paths, when rating information is unavailable in the user-item preference matrix. These metapaths cannot conform to the obtained recommendations as associated explanations. The research work discussed above in [23] provided the explanations to the product recommendations by constructing a knowledge graph from the probabilistic data model. This model is incomplete in terms of clarifying about the knowledge paths as recommendations with proper recommendation construct.

Recently, Belen et al. proposed [25] the utilization of Formal Concept Analysis (FCA) to generate explanations for the results of a recommender system. The limitation of using FCA to provide explanations is that there is much lower precision in the grouping of objects with shared properties [26] in constructing taxonomic relationships. This is because FCA does not support the sophisticated expression of concepts. In addition, they have worked on movie reviews with the ratings. The explanation of the movie recommendations based on ratings lead to biased explanations as the reviewers provide high ratings with the lack of negative measure.

The Web Ontology Language (OWL) [27] based development of aspect-based sentiment analysis ontology with a recommendation concept and the successive semantic web rules written on top of the ontology is offered as a solution for these problems. These rules help find the knowledge structure as explanations of the recommendations. This ontology also solves the limitations exhibited by the FCA model in terms of sufficient reasoning of the expressed concepts.

3.3 Motivation

E-commerce consumers prefer to view only those products that match their needs ideally during the search process while maintaining their autonomy of choice. This visibility of the preferred products results from the successful working of the recommender systems. The consumer satisfaction of the recommended products depends heavily on the amount of control these consumers have over the recommendation process. This substantially increases the appraisal of the recommendations by them.

However, today's recommender systems act as abstraction models; they are unable to provide any rationale for recommendations. The adoption of product recommendations by the consumers is only possible when the recommendation algorithms become comprehensive enough to contemplate the recommendations obtained by leveraging the user-generated reviews and provide the reasons as explanations in the intelligent manner.

The OWL-based approach to products recommendation explanations increases the adoptability of the recommendations in a trustworthy manner and better meets the principled requirements of future artificial intelligence.

3.4 Recommending Products Using Case-Based Reasoning

The pipeline of the product recommendations using case-based reasoning is illustrated in Figure 3.1.

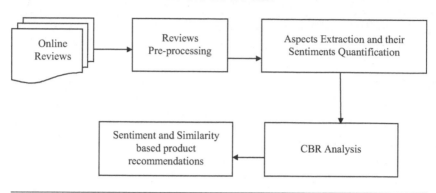

Figure 3.1   Pipeline of sentiment- and similarity-based products recommender system

3.4.1 Extracting Product Aspects

The task of aspect extraction from online reviews is performed by the implementation of three input channels to the CNN network. The network is first provided with two sets of vectors as input, and each with an input channel. The first input channel is fed with word embeddings and lexicon and the second input channel is fed with domain embeddings. These inputs are taken by both the channels and are submitted to the convolution layer. The convolution layer performs a convolution operation on these inputs. The convoluted intermediate is passed through Rectified Linear Unit (ReLU) activation function to generate feature maps. Full-length mathematical modelling of this (our) modified CNN model will be discussed in the forthcoming paper by Vidya et al. (manuscript in preparation).

The feature maps and the third input, namely the Part-of-Speech (POS) tokens, are now fed as inputs into the convolution layer present in the deep network of CNN. Once again, the convolution layer performs a convolution operation on these inputs. The convoluted intermediate is passed through Rectified Linear Unit (ReLU) activation function to generate feature maps. These feature maps are passed on to the pooling layer where the max pooling is performed on these feature maps. The output of the max pooling operation generates a higher representation of textual terms. These terms are fed as input to the fully connected layer in order to detect the potential product aspects. The output of multiple potential aspects is finally given to the Softmax classifier layer. This layer presents the highest probable term as the product aspect. In this way all the product aspects are extracted.

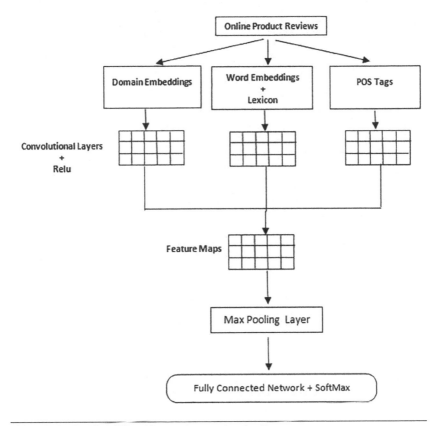

Figure 3.2 Modified CNN architecture for aspects extraction

The architecture of the modified CNN for the aspect extraction by the current researchers of this paper is shown in Figure 3.2.

The algorithm of the modified CNN model for aspects extraction is presented below.

INPUT: Pre-processed product reviews
OUTPUT: Extracted product aspects

ALGORITHM EXTRACT_PRODUCT_ASPECTS

1. **procedure** our_model (length, vocab_size)
2. repeat steps 3 to 8 for i, j, k, l, m, n←1 to 3 with varying kernel_size
3. input$_i$ ← Input(shape)

4. embedding$_j$ ← Embedding(vocab_size) (input$_i$)
5. conv$_k$ ← Conv1D (filters, kernel_size, activation) (embedding$_j$)
6. drop$_l$ ← Dropout (0.5) conv$_k$
7. pool$_m$ ← MaxPooling1D(pool_size)(drop$_l$)
8. flat$_n$ ← Flatten () (pool$_m$)
9. merge ← concatenate (flat$_1$, flat$_2$, flat$_3$)
10. dense ←Dense (10, activation ← 'relu') (merge)
11. outputs ←Dense (1, activation ← 'sigmoid') (dense)
12. model ←Model [input$_i$, outputs]
13. model.compile(loss, optimizer, metrics)
14. return outputs
15. end **procedure**

3.4.2 Evaluating Aspect Opinions and Sentiments

The adjectives, verbs and adverbs, as specified by Turney and Littmann [28]. The matched POS tags deemed as opinions are further evaluated for orientations by using the modified SO metric [17]. These orientations are counted for positive, negative and neutral aggregations. Finally, these aggregations are utilized in the Sentiment metric determined upon the aspect.

3.4.3 Generation of Base Cases and Query Case

The k-comparable product aspects as well as k-similar product aspects are under the same category are under the same category and are compared using their sentiments. The comparison is between the search query and the products available in the e-commerce products catalogue database. In CBR terminology, the search query is called query case and the database products are called base cases.

3.4.4 Recommending Products Using Sentiment and Similarity between Query Case and Base Cases

The query case and base cases are calculated for nearness by using Cosine Similarity [29]. The cosine angle between the query case and the base case is determined. When the angle is nearing 0° then the recommender system sorts these base cases in the increasing order of cosine angle towards 90° and shows them as recommendations to the user.

3.5 Explanations to the Product Recommendations by Ontological Reasoning

The generation of explanation (to the obtained recommendation) is the traversal of the path in the knowledge graph model rendered with the earlier extracted crucial pieces of knowledge. The relationships between the higher-level concepts and with the concepts of the extracted knowledge pieces help find the paths with required reasons to justify the recommendation. The pipeline of product recommendations explanation is illustrated in Figure 3.3.

This is achieved by first engineering the Explainable Product Reviews ontology (EPRO) in a top-down approach through incorporating the domain knowledge of product reviews with the help of concepts and properties between the concepts. The extracted crucial terms from the reviews are the instances to the earlier defined concepts. These instances are annotated with the ontology concepts and are useful in the downstream task of machine reasoning the recommendations by using the ontological knowledge model. The engineered ontology for the product reviews domain is shown in Figure 3.4.

This ontology also has the Recommendation concept. This is not visible in the above ontology visualization in Figure 3.3 as the main goal of this work is to find the name of the Recommendation entity instance during the introspection for explanations. The concepts in EPRO ontology is shown in Figure 3.5.

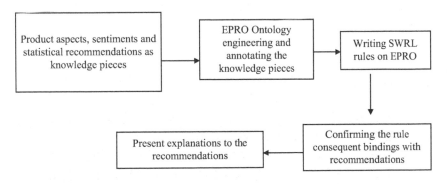

Figure 3.3 Pipeline of product recommendations explanations

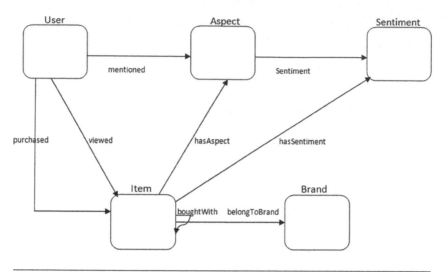

Figure 3.4 Visualization of explainable product reviews ontology

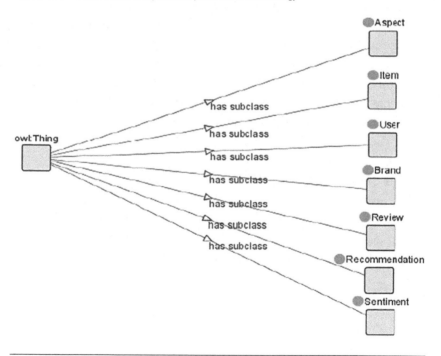

Figure 3.5 Concepts in EPRO ontology

Although the engineered EPRO ontology is developed with Description Logic (DL)-based rich semantics, it does not cover the full range of expressive possibilities so as to obtain the reasons for explaining the recommendations. The Semantic Web Rule Language (SWRL)-based rules allow the annotated data to be reasoned upon the engineered EPRO formal ontology in a consistent and logical manner, leading to the discovery of new relationships. The SWRL rules are built on top of the EPRO ontology. These rules facilitate the inclusion of more sophisticated rule-based constraints. The constraints involved in this research discover the relations among users, products, product aspects, their sentiments and the obtained recommendations in the form of explanations. The two SWRL rules are built to find out the explanation paths to the Recommendations. The rules are written below.

Rule Name 1: Aspect Embedded Product Recommendation Explanation
SWRL Rule 1: hasSentiment(?x, ?y) ∧ hasAspect(?x, ?z) ∧ hasSentiment(?z, ?y) → Recommendation(?x)

The corresponding natural language interpretation of the explanation is as follows:

[This product] is recommended because this product has [positive] sentiment and this product contains [aspect1, aspect2, aspect3…] as [positive] aspects which you might be interested in.

The graphic representation of SWRL rule 1 is shown in Figure 3.6.

Rule Name 2: User-Product Mapping Embedded Product Recommendation Explanation
SWRL Rule 2: User (? x) ∧ sqwrl:count(?x) ∧ purchased(?x, ?y) ∧ boughtWith(?y, ?z) ∧ viewed(?x, ?z) → Recommendation(?y)

The corresponding natural language interpretation of the explanation is as follows:

[This product] is recommended because [X number of users] had [purchased] this product earlier and this is [bought together] with [other product]. The [other product] you have [viewed] earlier.

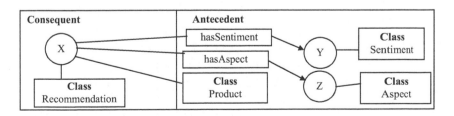

Figure 3.6 Graphic representation of SWRL rule1

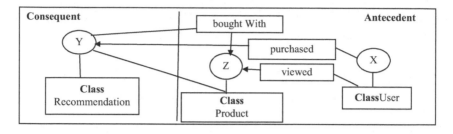

Figure 3.7 Graphic representation of SWRL rule2

The graphic representation of SWRL rule 2 is shown in Figure 3.7.

The pseudo-code of the explainable product recommendations is given below.

INPUT: The product to be searched
OUTPUT: Explainable product recommendations

RECOMMEND_PRODUCTS_WITH_EXPLANATION (ONTOLOGY O)

Start

1. Fetch the comparable products from the e-commerce product category database. The aspects of fetched products and the sought product are called as 'k-common aspects'.
2. For all of the k-common aspects, compute the sentiment as follows;

$$\text{Sentiment}\left(A_j, P_i\right) = \frac{\text{count}(\text{Pos}(A_j, P_i)) - \text{count}(\text{Neg}(A_j, P_i))}{\text{count}\left(\text{Pos}\left(A_j, P_i\right)\right) + \text{count}\left(\text{Neg}\left(A_j, P_i\right)\right) + \text{count}\left(\text{Neu}\left(A_j, P_i\right)\right)}$$

3. Map Positive = [0.001 ... 1] and Negative = [-1 ... -0.001].
4. Determine the relevant products to the searched product using aspects sentiments in cosine similarity.
5. Arrange the products in the descending order of cosine similarity.

6. Annotate products, aspects, sentiments as individuals to the constructs of O.
7. Build the following SWRL rules on top of O.
   7.1 **hasSentiment(?x, ?y) ∧ hasAspect(?x, ?z) ∧ hasSentiment(?z, ?y) → Recommendation(?x)**
   7.2 **User (? x) ∧ sqwrl:count(?x) ∧ purchased(?x, ?y) ∧ boughtWith(?y, ?z) ∧ viewed(?x, ?z) → Recommendation(?y)**
8. Debug the SWRL rules for explainable product recommendations.
9. Reasoner selects that Recommendation individual for the rule consequent which satisfies the conjunction of DL-safe triple patterns on the relation constructs from O in the rule antecedent.
10. Confirm the bindings of the rule consequent with the statistical recommendation.
11. Recommend products with explanations.

End

The Antecedent or Left-Hand Side (LHS) of the two SWRL rules is deemed to be the knowledge paths to explain the recommendations as the inferred instances are the materialized knowledge from the EPRO ontology. The Consequent or Right-Hand Side (RHS) of the two rules are the inferred instances of the Recommendation concept satisfying the materialized path of LHS. The debugged SWRL rules with logical consequences as computed by the reasoner are specified separately in Appendix A.

3.6 Experimental Setup and Results Discussion

This task draws on five categories of product reviews from Amazon: smartphones, laptops, cameras, books and wristwatches. In every product category, a total of 100 products are considered. Table 3.1 presents the details of the datasets used for this experiment.

TABLE 3.1   Details of the Considered Dataset

DOCUMENT CHARACTERISTICS	STATS
Total review documents	4,13,841 (129MB)
Number of minimum sentences per review	1
Number of maximum sentences per review	42
Average number of reviews by customers	5.79
Average number of reviews on the product	48.44

TABLE 3.2   Experimental Setup

PARAMETER/TECHNIQUE/FUNCTION	VALUE
Filter size	(3,4,5)
Number of feature maps	100
Function used for Pooling	MAX
Regularization Technique used (To prevent over-fitting)	Drop Out
Activation Function used	ReLU
Number of epochs	10

The domain embeddings are trained on the collected Amazon reviews. The collected reviews fall under the electronic goods domain. The word embeddings utilized in this work is the Google's word2vec embeddings Google News corpus. The pre-processing of data is carried out by removing stop words and non-English words. POS tagging is then performed on the obtained set of words. The experimental setup comprising of various settings for the modified CNN model is tabulated in Table 3.2.

The machine used to implement our modified CNN model using a Python framework is configured with 8GB RAM, an i5 multi-core processor and the HDD of 1TB. The modified CNN with these carefully configured settings has achieved an accuracy of 81.90%. The statement "Bigger data better than smart algorithms" by Pedro Domingos in [30] is well suited to our modified CNN model, which is clear from the aspects of model accuracy with considerable size of the reviews collection. The extracted aspects from the electronic goods domain (two product results are shown here) are tabulated in Table 3.3.

The verbs and adverbs present in the review sentences are analyzed for opinions. The effect of verbs on the valence of opinion of

TABLE 3.3 Extracted Aspects by Our Modified CNN Model

PRODUCT	ASPECTS EXTRACTED
Wristwatches	Wristwatch, band, wrist, face, watches, strap, size, item, order, ordered, refund, quality, made, good, plastic, solid, sturdy, product, construction
Smartphones	headset, Bluetooth, belt, sound, quality, volume, noise, phone, clip, screen, headsets, ear case, product, read, cover, plastic, leather, hear

TABLE 3.4 Count of Aspect: Opinion Pairs with Verbs

PRODUCT CATEGORY/ASPECT, OPINION PAIRS (VERBS)	POSITIVE/ NEGATIVE COUNT	TOTAL COUNT OF VERBS	% OF IMPLIED POSITIVE/ NEGATIVE OPINIONS
Wristwatches/386	199+/187-	3588	5.5% / 5.2%
Books/678	358+/320-	4693	7.6% / 6.8%
Laptops/797	436+/361-	4306	10.8% / 9.3%
Smartphones/529	345+/184-	5211	6.7% / 3.8%
Cameras/765	447+/318-	4237	10.3% / 7.3%

the product aspects is carried out to find out whether verbs have any impact on the opinion analysis. The amount of aspect, opinion pairs identified using verbs and the positive or negative implications are tabulated in Table 3.4.

The percentage of implied positive opinions and negative opinions identified by the verbs tabulated above provides the information that verb has good influence on the opinion analysis.

The effect of adverbs on the valence of opinion of the product aspects is carried out to find out whether adverbs have any impact on the opinion analysis. The amount of aspect, opinion pairs identified using adverbs and the positive or negative implications are tabulated in Table 3.5.

The percentage of implied positive opinions and negative opinions identified by the adverbs tabulated above provides the information that adverbs have good influence on the opinion analysis.

The sentiments are calculated by using the formula specified in the algorithm earlier in Section 3.4. The value of 'k' in k-common aspects

TABLE 3.5 Count of Aspect: Opinion Paris from Adverbs

PRODUCT CATEGORY/NO. OF ASPECT, OPINION PAIRS (ADVERBS)	POSITIVE/ NEGATIVE COUNT	TOTAL COUNT OF ADVERBS	% OF IMPLIED POSITIVE/ NEGATIVE OPINIONS
Wristwatches/422	396+/26−	2669	14.7% / 1.2%
Books/697	483+/214−	3958	12.2% / 5.4%
Laptops/809	669+/140−	2030	33.0% / 6.8%
Smartphones/614	452+/162−	4366	10.3% / 3.7%
Cameras/786	704+/82−	3852	18.2% / 2.1%

TABLE 3.6 Variation in k Value

PRODUCT CATEGORY	'k'
Wristwatches	8
Books	17
Laptops	23
Smartphones	21
Cameras	13

differs based on the count of common aspects determined from the class of comparable products. The variability in the value of 'k' in the five categories are tabulated in Table 3.6.

The algorithm calculates sentiments for all of the k-common aspects of 150 products. In order to evaluate the sentiments of the k-common aspects for recommending products, Cosine Similarity is considered. The number of comparable products recommended across the five product categories is tabulated in Table 3.7.

The observation from Table 3.7 is that the recommendation accuracy rises uniformly when the value of k is increasing. This is observed when aspect sentiments are utilized in ascertaining the recommendations.

The sample set of product recommendations by analyzing 31,437 reviews with 15 products is tabulated in Table 3.8. The recommendations are based on the User-Product Mapping information of the product aspects.

It is observed that an Area under ROC curve (AUC) of 91% is achieved across all five product categories. From this result, it is concluded that the sentiment and similarity-based product recommendations provide user relevant recommendations.

TABLE 3.7 Percentage of Products Recommended across Five Categories

PRODUCT CATEGORY	'k' VALUE	ASPECT SENTIMENTS-BASED PRODUCT RECOMMENDATIONS
		RECOMMENDATION ACCURACY (%)
Wristwatches	k=5	59.7
	k=8	79.4
Books	k=5	23.1
	k=10	53.6
	k=15	73.4
	k=17	86.5
Laptops	k=5	32.9
	k=10	42.7
	k=15	55.5
	k=20	69.8
	k=23	83.2
Smartphones	k=5	36.4
	k=10	66.2
	k=15	76.3
	k=20	91
	k=21	93.4
Cameras	k=5	42.9
	k=10	73.6
	k=13	89.7

TABLE 3.8 Sample Set of Product Recommendations w.r.t. User Product Mapping Embedding

PRODUCT CATEGORY/SEARCHED PRODUCT/K VALUE	PRODUCT RECOMMENDATIONS ORDER - (PRODUCT1, PRODUCT2)
Wrist Watches / Fossil Men's Gen 4 Explorist / 8	Fossil Neutra Hybrid Analog Dial Men's watch, Fossil Machine Analog Black Dial men's watch
Books / Deep Learning with Python / 15	Thinking Machines/Tensorflow for machine intelligence
Laptops / Dell Inspiron Core i3 6th Generation / 15	Dell Inspiron 15 3567, **NIL**
Smart phones / Apple iPhone 6s plus / 15	iPhone 7 plus, iPhone case
Cameras / Nikon Coolpix A10 point and shoot / 15	Nikon Coolpix A10 Point and Shoot Digital Camera (Silver) with Memory Card and Camera Case, Sony DSC – H400 point and shoot

The engineered EPRO ontology is primarily evaluated upon two major criteria namely consistency of the ontology and the completeness of the ontology [31]. The logical consequences obtained from the defined SWRL rules are found to be non-contradictory and is inferentially consistent. In addition, the reasoned consequences from the defined SWRL rules revealed that the Recommendation knowledge which is not related explicitly with other concepts and properties are still inferred. This proves the completeness of the ontology.

The statistical evaluation of SWRL rules-based explanations to recommendations is carried out offline. The evaluation measure is the percentage of product recommendations that are explained by the proposed explanation model. Mean Explainability Precision (MEP) and Mean Explainability Recall (MER) are used to evaluate the top-n recommendations with their explanations. The product reviews considered for the purpose of evaluation is from Amazon's smartphones category. The value of n is increased in the interval of 10 up to the maximum of 50. The MEP@50 and MER@50 with the proposed model is tabulated below in Table 3.9.

The MEP and MER values obtained by using the proposed method are compared against the baselines, namely Item based Explainable Matrix Factorization [24] (EMF_{IB}) and User-based Explainable Matrix Factorization [24] (EMF_{UB}) in the above table. It is observed that the proposed method outperformed other techniques in terms of MEP and MER. This is because when there is an increase in the n value, it is observed that the MEP values for EMF_{UB} and EMF_{IB} are increasing and the MER values for EMF_{UB} and EMF_{IB} are decreasing. The decreasing trend is because the explanations that are provided on the products are less in number. The results are visualized in Figures 3.8 and 3.9.

It is also observed with the proposed method, when the n value is increasing for recommending top-n products, the percentages of

TABLE 3.9 MEP- and MER-Based Evaluation of Product Recommendations Justification

	MEP@50				MER@50		
N	EMFIB	EMFUB	PROPOSED MODEL	N	EMFIB	EMFUB	PROPOSED MODEL
10	0.5088	0.8033	0.8764	10	0.0125	0.0119	0.0368
20	0.5092	0.8952	0.9159	20	0.0112	0.0106	0.0323
30	0.5092	0.8965	0.9193	30	0.0108	0.0103	0.0284
40	0.5091	0.8973	0.9279	40	0.0105	0.0098	0.0189
50	0.5091	0.8984	0.9362	50	0.0103	0.0094	0.0117

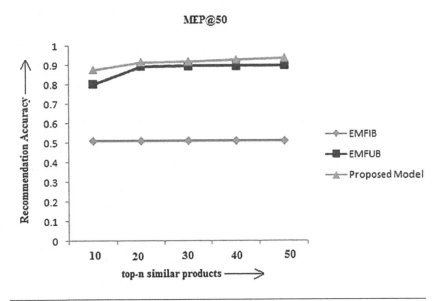

Figure 3.8   MEP@50 comparative analysis

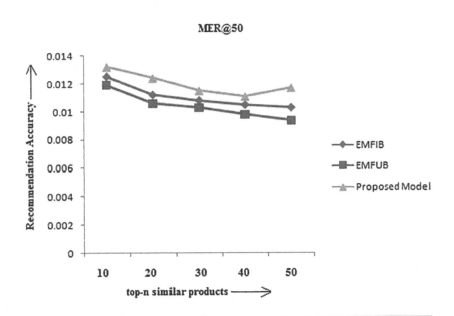

Figure 3.9   MER@50 comparative analysis

product recommendations that are explained are also increasing. This is because of the average that is taken across the top-n recommendations.

The comparative analysis of the engineered EPRO ontology-based explainable recommendations with FCA lattice-based explainable recommendations [25] on the basis of common metrics to both of them are shown in Figure 3.10.

The number of nodes is compared, based on the number of items (movies in FCA and products in proposed work) used to generate the model. The minimum number of nodes with the number of individual movies in FCA is 28 and with the number of individuals the minimum number of nodes in the proposed ontology is 32. This specifies that when the size of the ontology in terms of nodes grows rapidly the finding of the explanation paths happens in the bounded manner.

On the other hand, in the FCA-based approach it is specified that the optimal size to compute an e-commerce user lattice is 20. The growth of the lattice is specified to be decreasing thereafter. This clearly indicates that the possibility of finding the tailored explanations happens incompletely.

Then the width of the models is compared. When it comes to width metric in the FCA model, the minimum number of nodes in a lattice level is 8. It is understood that the larger the user preferences, the more the heterogeneity of the user profiles, thereby making the

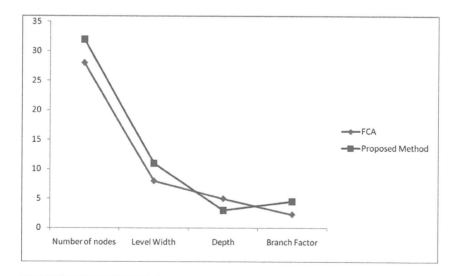

Figure 3.10   Common metrics for comparison between FCA and proposed OWL models

lattice destabilized. This leads to increase of computational time in finding the actual explanation paths. This problem was specified in [32]. In addition, the destabilized FCA lattice generates those reasons as explanations that contain in them the irrelevant conceptual items.

On the other hand, in the proposed ontology model the minimum number of nodes at a level (width) is 11. These are found to be the annotated individuals to two previous level concepts. This specifies that when the number of individuals increases there is no point of destabilized ontology being formed as the number of concepts and the properties are fixed. Ontology gets destabilized when it is under the process of evolution [33]. As the ontology engineered for this research is a stable resource, the explanations are found from the ontology are relevant and in less computational time.

Moving on, the depth of the models is compared. The FCA model has a minimum depth of 5 in a user profile. This specifies that as the number of attributes of the item increases, the depth of the lattice increases making the lattice more homogeneous. This metric as it is understood as homogeneous the possibility of generating explanation path is better. However, the analysis of the width/depth ratio on the basis of number of items in the lattice reduces the chances of finding the relevant reasons as explanations to the recommendations by a fair amount.

The minimum depth of the proposed ontology model is 3. This clearly specifies that the path from the individual residing at the bottom of the ontology tree makes itself readily available for communicating with the properties among the other individuals. The width/depth ratio-based analysis here increases the likelihood of finding the relevant reasons for explanations to the recommendations. The likelihood is the conjunctions specified between the properties in SWRL rules. This means that there must be an individual that is associated with those two properties. It is observed that there is a common individual between the two properties in the association.

Finally, a comparison is made of the branch factor of the models. The branch factor of FCA is 2.3. This specifies that finding the explanation path is easy as there is a smaller number of intermediate nodes containing shared attributes. However, this is again at the cost of depth versus width trade-off. In our comparison, the FCA model has less chance of selecting the appropriate item as recommendation outcome.

The proposed ontology has a branch factor of 4.5. This specifies that there are a higher number of intermediate nodes containing shared properties. It is observed from the ontology model that these intermediate nodes are the child nodes that became associated with other node properties. The lesser value of depth versus width trade-off in our ontology model increases the chances of selecting the appropriate item as recommendation outcome.

Therefore, the OWL-based explaining of the product recommendations make the users aware of why they are seeing such recommendations and helps to improve the effectiveness, efficiency, persuasiveness and user satisfaction of recommender systems over FCA based explaining the product recommendations.

The advantages these results bring to the e-commerce system are that the product aspects and their sentiments help the customers to make up their mind in purchasing the product in the wiser manner.

3.7 Conclusion and Future Scope

The task of machine reasoning the recommendations by using OWL-based SWRL rules analysis was carried out successfully. The product aspects were extracted by implementing the modified CNN model with three input channels. The co-related opinions were analyzed and are aggregated for sentiments. These similar aspects across CBR query case and base cases were compared and are presented with recommendations. The above-mentioned recommendations are persuasive to users by making the engineered ontology with SWRL rules reason the knowledge paths as explanations. Finally, these OWL model-based explanations are compared with FCA model-based explanations with four evaluation metrics which revealed that OWL-based explanations are complete and DL-safe as compared to FCA based explanations.

In future, this recommender system is integrated with the user's social profiles in order to better infer the user preferences from other channels and make the explanations to the recommendations more transparent.

SWRL Debugger

▼ X

⤢ Clear output

Name	Value
▲ ✦ Rule	If an user is-purchased a thing(1) and the thing(1) is-bought-with a thing(2) and the user is-viewed the thing(2) then the thing(1) is a recommendation.
user	User-A
thing(1)	Iphone
thing(2)	Iphone-Case<http://www.owl-ontologies.com/EPRO.o#>
▲ ✦ Rule	If a thing(1) has-sentiment a thing(2) and the thing(1) has-aspect a thing(3) and the thing(3) has-sentiment the thing(2) then the thing(1) is a recommendation.
thing(1)	Iphone
thing(2)	Positive
thing(3)	Iphone-O-S<http://www.owl-ontologies.com/EPRO.o#>

Figure 3.A.1   Debugged SWRL rules as explanations to the recommendations

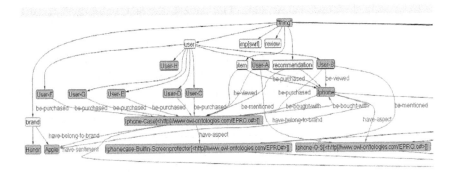

Figure 3.A.2   Part of the materialized graph showing the knowledge path for Recommendation

Appendix 3.A

The debugged SWRL rules of the product reviews collection are illustrated in Figure 3.A.1 and Figure 3.A.2.

The materialized graph as the knowledge path after debugging the SWRL rules are shown as part of the complete graph emphasizing the Recommendation instance in Figure 3.A.2.

References

[1] Dong, R., Schaal, M., O'Mahony, M.P., McCarthy, K., & Smyth, B. (2013). Opinionated Product Recommendation. In: Delany S.J., & Ontañón S. (Eds.) *Case-Based Reasoning Research and Development. ICCBR 2013. Lecture Notes in Computer Science*, vol 7969. Springer, Berlin, Heidelberg.

[2] Xiao, J., Ye, H., He, X., Zhang, H., Wu, F., & Chua, T. S. (2017). Attentional factorization machines: Learning the weight of feature interactions via attention networks. arXiv preprint arXiv:1708.04617.

[3] Seo, S., Huang, J., Yang, H., & Liu, Y. (2017, August). *Interpretable convolutional neural networks with dual local and global attention for review rating prediction.* In *Proceedings of the Eleventh ACM Conference on Recommender Systems* (pp. 297–305).

[4] Schafer, J. B., Konstan, J. A., & Riedl, J. (1999, December). *Recommender systems in e-commerce.* In 1st *ACM Conference on Electronic Commerce, EC 1999* (pp. 158–166).

[5] Zhang, Y., Lai, G., Zhang, M., Zhang, Y., Liu, Y., & Ma, S. (2014, July). *Explicit factor models for explainable recommendation based on phrase-level sentiment analysis.* In *Proceedings of the 37th international ACM SIGIR conference on Research & development in information retrieval* (pp. 83–92).

[6] Huang, J., Zhao, W. X., Dou, H., Wen, J. R., & Chang, E. Y. (2018). Improving Sequential Recommendation with Knowledge-Enhanced Memory Networks.

[7] Zhang, Y., & Chen, X. (2018). Explainable recommendation: A survey and new perspectives. arXiv preprint arXiv:1804.11192.

[8] Schank, R. C., Kass, A., & Riesbeck, C. K. (Eds.). (2014). *Inside case-based explanation.* Psychology Press.

[9] Dong, R., O'Mahony, M. P., Schaal, M., McCarthy, K., & Smyth, B. (2016). Combining similarity and sentiment in opinion mining for product recommendation. *Journal of Intelligent Information Systems,* 46(2), 285–312.

[10] Poria, S., Cambria, E., & Gelbukh, A. (2016). Aspect extraction for opinion mining with a deep convolutional neural network. *Knowledge-Based Systems,* 108, 42–49.

[11] Ye, H., Yan, Z., Luo, Z., & Chao, W. (2017, May). *Dependency-tree based convolutional neural networks for aspect term extraction.* In *Pacific-Asia Conference on Knowledge Discovery and Data Mining* (pp. 350–362). Springer, Cham.

[12] Luo, H., Li, T., Liu, B., Wang, B., & Unger, H. (2019). Improving aspect term extraction with bidirectional dependency tree representation. *IEEE/ACM Transactions on Audio, Speech, and Language Processing,* 27(7), 1201–1212.

[13] Xu, H., Liu, B., Shu, L., & Yu, P. S. (2018). Double embeddings and cnn-based sequence labeling for aspect extraction. arXiv preprint arXiv:1805.04601.

[14] Da'u, A., & Salim, N. (2019). Aspect extraction on user textual reviews using multi-channel convolutional neural network. *Peer Journal of Computer Science,* 5, e191.

[15] Ojokoh, B. A., & Kayode, O. (2012). A feature-opinion extraction approach to opinion mining. *Journal of Web Engineering,* 11(1), 51.

[16] Santosh, D. T., Babu, K. S., Prasad, S. D. V., & Vivekananda, A. (2016). Opinion Mining of Online Product Reviews from Traditional LDA Topic Clusters using Feature Ontology Tree and Sentiwordnet.

[17] Kumar, K. R., Santosh, D. T., & Vardhan, B. V. (2019). Determining the semantic orientation of opinion words using typed dependencies for opinion word senses and Senti Word Net scores from online product reviews. *International Journal of Knowledge and Web Intelligence*, 6(2), 89–105.

[18] Lakkaraju, H., Socher, R., & Manning, C. (2014, January). *Aspect specific sentiment analysis using hierarchical deep learning*. In *NIPS Workshop on Deep Learning and Representation Learning* (pp. 1–9).

[19] Gu, X., Gu, Y., & Wu, H. (2017). Cascaded convolutional neural networks for aspect-based opinion summary. *Neural Processing Letters*, 46(2), 581–594.

[20] Luo, H., Li, T., Liu, B., & Zhang, J. (2019, July). *DOER: Dual cross-shared RNN for aspect term-polarity co-extraction*. In *Proceedings of the 57th Annual Meeting of the Association for Computational Linguistics* (pp. 591–601).

[21] Teja Santosh, D., Kakulapati, V., & Basavaraju, K. (2020). Ontology-based sentimental knowledge in predicting the product recommendations: A data science approach. *Journal of Discrete Mathematical Sciences and Cryptography*, 23(1), 1–18.

[22] Catherine, R., Mazaitis, K., Eskenazi, M., & Cohen, W. (2017). Explainable entity-based recommendations with knowledge graphs. arXiv preprint arXiv:1707.05254.

[23] Ai, Q., Azizi, V., Chen, X., & Zhang, Y. (2018). Learning heterogeneous knowledge base embeddings for explainable recommendation. *Algorithms*, 11(9), 137.

[24] Abdollahi, B., & Nasraoui, O. (2017, August). *Using explainability for constrained matrix factorization*. In *Proceedings of the Eleventh ACM Conference on Recommender Systems* (pp. 79–83). ACM.

[25] Diaz-Agudo, B., Caro-Martinez, M., Recio-Garcia, J. A., Jorro-Aragoneses, J., & Jimenez-Diaz, G. (2019, September). *Explanation of Recommenders Using Formal Concept Analysis*. In *International Conference on Case-Based Reasoning* (pp. 33–48). Springer, Cham.

[26] Asim, M. N., Wasim, M., Khan, M. U. G., Mahmood, W., & Abbasi, H. M. (2018). A survey of ontology learning techniques and applications. *Database*, 2018.

[27] Antoniou, G., & Van Harmelen, F. (2004). Web ontology language: Owl. In *Handbook on ontologies* (pp. 67–92). Springer, Berlin, Heidelberg.

[28] Turney, P. D., & Littman, M. L. (2003). Measuring praise and criticism: Inference of semantic orientation from association. *ACM Transactions on Information Systems (TOIS)*, 21(4), 315–346.

[29] Huang, A. (2008, April). *Similarity measures for text document clustering*. In *Proceedings of the sixth new zealand computer science research student conference (NZCSRSC 2008)*, Christchurch, New Zealand (Vol. 4, pp. 9–56).

[30] Domingos, P. (2012). A few useful things to know about machine learning. *Communications of the ACM*, 55(10), 78–87.

[31] Fitzgerald, S. J., & Wiggins, B. (2010). Staab, S., Studer, R. (Eds.), Handbook on Ontologies, Series: International Handbooks on Information Systems, Vol. XIX (2009). 811 p., 121 illus., Hardcover£ 164, ISBN: 978-3-540-70999-2. *International Journal of Information Management*, 30(1), 98–100.

[32] Codocedo, V., Taramasco, C., & Astudillo, H. (2011, October). Cheating to achieve formal concept analysis over a large formal context.

[33] Orme, A. M., Yao, H., & Etzkorn, L. H. (2007). Indicating ontology data quality, stability, and completeness throughout ontology evolution. *Journal of Software Maintenance and Evolution: Research and Practice*, 19(1), 49–75.

4

MODEL-BASED FILTERING SYSTEMS USING A LATENT-FACTOR TECHNIQUE

ALEENA MISHRA, MAHENDRA KUMAR GOURISARIA, PALAK GUPTA, AND SUDHANSU SHEKHAR PATRA

KIIT Deemed to be University, Bhubaneswar, Odisha, India

LALBIHARI BARIK

King Abdul Aziz University, Jeddah, Kingdom of Saudi Arabia

Contents

4.1 Introduction

Collaborating filtering methods are famous algorithms for recommender systems. Thence the products suggested to clients are controlled by looking over the networks. Collaborating filtering methods are divided into two types: (1) the Model-based method; and (2) the Memory-based method. A memory-based algorithm loads the whole database into system memory and makes forecasts for suggestions dependent on such online memory databases. Model-based algorithms attempt to pack gigantic databases into a model and performs proposal task by applying the recommendation mechanism. A model-based collaborating filter can answer the client's solicitation instantaneously. This chapter reviews regular methods for executing model-based recommender systems. Through the use of some methods of machine learning such as supervised or unsupervised mechanisms, a simplified form of data is generated in model-based approaches. The instances of classical machine learning methods employed in this approach include Bayes classifiers, rule-based techniques, regression models, SVMs, decision trees, as well as neural networks, and so on (Aggarwal 2015). An unsupervised model, such as the k-nearest neighbor classifier, can be summed up to neighborhood-based frameworks, all those classical models given above can also be summed up to the collaborating filtering scenario because these classical problems are also general events of collaborating filtering problem (Charu 2016; Basilico and Hofmann 2004). Our main focus of the discussion is on the latent factor model. First, we briefly discuss four types of the common methodology of the model-based model, these are clustering, classification, latent model (Kuanr and Mohanty 2018), Markov decision process. Second, we discuss the geometrical intuition of latent factor, factorization matrix, integrating factorization, and neighborhood model followed by integrating neighborhood and latent factor model (Kuanr, Mohanty, Sahoo, and Rath 2018).

4.2 Types of Model-Based Filtering

4.2.1 Clustering Collaborating Filtering

Clustering CF (Sarwar et al. 2001) depends on the intuition that clients in a similar gathering have common traits. Accordingly, under such systems bunch clients together into groups, which are characterized as a lot of comparable clients (Garanayak, Mohanty, Jagadev, and Sahoo 2019). Assume every client is spoken to as a rating vector indicated by $u_i = (m_{i1}, m_{i2}, m_{i3}, \dots, m_{in})$. The separation between the two users is the difference measure between two clients. In making the calculations, Minkowski distance, Euclidian distance, or Manhattan distance can be used as indicated below.

$$\text{Distance}_{\text{Manhattan}}\left(u_1, u_2\right) = \sum_{j} \left| m_{1j} - m_{2j} \right|$$

$$\text{Distance}_{\text{Minkowski}}\left(u_1, u_2\right) = \sqrt[a]{\sum_{j} \left(m_{1j} - m_{2j}\right)^2}$$

$$\text{Distance}_{\text{Euclidean}}\left(u_1, u_2\right) = \sqrt{\sum_{j} \left(m_{1j} - m_{2j}\right)^2}$$

Clustering collaborating filtering experiences the issue of meager rating matrix where there are many missing qualities; this can cause clustering calculations to be loose. To take care of this issue, Sarwar et al. (2001) proposed an inventive clustering collaborating filtering which directly bunches things dependent upon which clients rate such things and afterwards utilizes the grouping of items to help bunch clients. Their technique is a formal statistical model.

4.2.2 Classification Collaborating Filtering

In classification collaborating filtering, we classify users according to their ratings. Let us represent the user as a rating vector which can be denoted as $u_i = (m_{i1}, m_{i3}, \dots, m_{in})$ (Thi et al. 2010). Each rating values m_{ij} are integer, which range from r_1 to r_m. Let us consider a 5-value rating system, in which $r_i = i$, where $i \in [1, 5]$ (Thi et al. 2010). If an item is given a rating of 5 by the user that means he/she has the strongest possible preference. By contrast, if an item is rated as 1 by the user that means he/she prefers the least such thing. In this manner we can classify users based on their ratings.

4.2.3 Markov Decision Process (MDP)-Based CF

The recommendation can be considered as a consecutive procedure, including numerous stages (Shani et al. 2005; Hofmann and Puzicha 1999). At each stage a rundown of things which is resolved dependent on the last client's appraising that is suggested to the client. To satisfy the user's interest, the suggestion task is the best activity that the recommender framework must do at a concrete stage. The suggestion turns into the way toward settling on the choice to pick the best activity. This approach of giving suggestions or recommendations is proposed (Shani et al. 2005; Hofmann and Puzicha 1999).

4.2.4 Latent Factor

One of the cutting-edge technologies in terms of recommender systems is the latent factor model. To fill in the missing entries, such models use some notable dimensional reduction mechanism. In the area of data analytics, the method of dimensional reduction is used widely to define fundamental information in a limited number of dimensions (Agarwal and Chen 2009). To maintain the correlation in pairs, dimensional reduction flips the coordinate structure. This is the general concept behind the dimensional reduction. In one shoot to appraise the whole information matrix, matrix factorization give a slick way to use all rows and columns relationship. This advancement of the methodology is one reason to make latent factor models a cutting edge in model-based collaborating filtering.

4.3 The Geometrical Intuition Behind the Latent Factor Model

In order that the reader should gain a good grasp of the geometrical intuition for the latent factor, it will be explained below. Let's take three things in an appraisal or rating matrix, where we can find the positive correlation among these three things (Charu 2016).

Let's take three movies, *Spartacus, Nero, and Gladiator,* for a movie rating scenario. The ratings lie in the interval of [-1 and 1] and it is assumed that these are continuous values. From Figure 4.1, we can see that the 3-dimensional scatter plot of rating is arranged in a 1-dimensional line as they are positively correlated. After expelling the outliers, the information is for the most part orchestrated alongside a 1-dimensional line, which implies that the information matrix may have a rank 1. From Figure 4.1, we can see that the latent vector goes through the focal point of the information and is lined up with the lengthened data dispersion. To find the projection of information along this line, some strategies of dimensionality reductions, such as (mean-focused) singular value decomposition (SVD) and principal components analysis (PCA), are used. We can represent the information of an m × n matrix having a rank p in a p-dimensional hyperplane, as shown in Figure 4.1.

If the p-dimensional hyperplane is aware to us for whatever length of time, we can easily estimate the missing values of users' ratings. On account of Figure 4.1, just one rating should be indicated to decide the other two evaluations, because 1 is the rank of the matrix when we have done with commotion expulsion. Let's take an instance where we fix the appraisal for *Gladiator* with a value of 0.5; accordingly, the rating of *Spartacus* and *Nero* can be evaluated because the rating of *Gladiator* is fixed at the meeting of the axis-parallel hyperplane with the 1-dimensional latent vector (in Figure 4.1) to 0.5. To determine

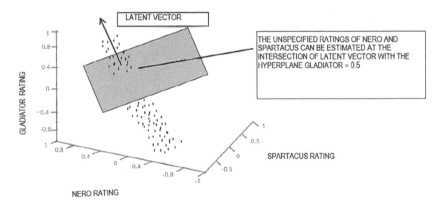

Figure 4.1   In this figure, we fix the rating value of *Gladiator* at 0.5 which can help in computing missing data for other users.

the unknown records, dimensional reduction methods such as single value decomposition use to control the layoffs and correlation.

4.4 Some Fundamental Matrix Factorization Concepts

The rating matrix M of dimension a × b can be written as:

$$M \approx XY^T \tag{4.1}$$

where matrix X is of order a × k and matrix Y is of order b × k. This is the approximate factorization of matrix M. We can use this concept for matrix factorization afterwards.

Related terms:

Latent vector: refers to every column of X or Y
Latent factor: refers to each row of X or Y
User factor: refers to i^{th} row x_i of X that includes k items that analogous to the similar users i towards the k concepts in the rating matrix (Charu 2016).
Item factor: refers to i^{th} row of y_i of Y that includes the similarity of ith item towards the k concepts.

The history and romance movies ratings are contained by a 2-dimensional vector m_{ij}. The item factor includes the similarity of the item of the two classes of movies, as shown in Figure 4.2. We can take the dot product of ith user factor and j^{th} item factor to express each rating m_{ij} in M according to the Equation (4.1). So we can write it as:

$$m_{ij} \approx x_i \cdot y_j \tag{4.2}$$

If latent factors $x_i = (x_{i1} \ldots x_{ik})$, $y_j = (y_{j1} \ldots y_{jk})$ give the similarities for users of the k difference concepts, then m_{ij} can be expressed as:

$$m_{ij} \approx \sum_{d=1}^{k} x_{id} \cdot y_{id} \tag{4.3}$$

$$= \sum_{d=1}^{k} \left(\text{affinity of client i to concept d} \right) \times$$

$$\left(\text{affinity of product j to concept d} \right)$$

If we express Figure 4.2 according to Equation (4.3), then it would be as given below:

$$m_{ij} \approx \left(\text{affinity of client i to romance} \right) \times \left(\text{affinity of product j to romance} \right)$$

$$+ \left(\text{affinity of client i to history} \right) \times \left(\text{affinity of product j to history} \right)$$

Most of the time the latent vector becomes a self-assertive vector with negative and positive quantities; as a result we find difficulties in giving it semantic translation. Figure 4.1 is the representation of a geometric correlation pattern, which can also be represented by it as a presiding correlation design in the ratings matrix. By using a specific type of matrix

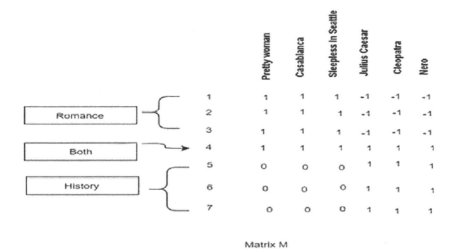

	Pretty woman	Casablanca	Sleepless In Seattle	Julius Caesar	Cleopatra	Nero
1	1	1	1	-1	-1	-1
2	1	1	1	-1	-1	-1
3	1	1	1	-1	-1	-1
4	1	1	1	1	1	1
5	0	0	0	1	1	1
6	0	0	0	1	1	1
7	0	0	0	1	1	1

Romance → rows 1–3; Both → row 4; History → rows 5–7

Matrix M

Figure 4.2 Resultant Matrix M, which is the product of two matrix X (here in our example, its row contains no of users and columns contains two genres of movies i.e. Romance and History) and Y (here in our example, its row contains two genres of movies i.e. Romance and History and columns contains six movie names: *Pretty Women, Casablanca, Sleepless in Seattle, Julius Caesar, Cleopatra, and Nero*).

factorization method such as non-negative matrix factorization, we can accomplish more noteworthy interpretability in the latent vectors.

4.4.1 Unconstrained Matrix Factorization

The factor matrices (X and Y) in which no constraints are imposed are known as unconstrained matrix factorization. This is the most essential type of matrix factorization. It is technically incorrect to say that singular value decomposition is the unconstrained matrix factorization as recommended by much of the existing literature. The columns of X and Y should be orthogonal in the case of single value decomposition. But the above conflict is so ubiquitous aming some recommendation researchers which causes perplexion amongst some beginners in this area.

We first discuss the issue of factorizing fully specified matrices; then we go on to the factorization of incomplete matrices. How might we decide the factor matrices X and Y, with the goal that the fully specified matrix M equals XY^T as intently as could reasonably be expected? In order to achieve this goal, one can figure an enhancement issue regarding the networks X and Y as given below:

$$\text{Minimize } J = \|M - XY^T\|^2 \tag{4.4}$$

subject to: No constraints on X and Y (Charu 2016)

The squared Frobenius norm of the matrix is represented as $\|.\|^2$, which is equivalent to the aggregation of the matrix elements, which is known as objective function. The set D of perceived client-things pairs is characterized as follows:

$$D = \left\{ (i,j) : m_{ij} \text{ is observed} \right\}$$

where
    (i, j) is the set of client-item pairs in M,
    i is the index of a client which belongs to {1 …. a} and
    j is the index of a thing which belongs to {1 … b}

All the items in M can be anticipated. If the incomplete matrix M can be factorized somehow as the product of fully specified matrices XY^T then the formula can be formulated as:

$$\bar{m}_{ij} = \sum_{d=1}^{k} x_{id} \cdot y_{jd} \qquad (4.5)$$

Here $\bar{m}$ is the predicted value. Error e_{ij} can be defined as the difference between the observed quantity and predicted quantity which can be formulated as:

$$\text{Minimise } J = \frac{1}{2} \sum_{(i,j)\in D} e_{ij}^2 = \frac{1}{2} \sum_{(i,j)\in D} \left(m_{ij} - \sum_{d=1}^{k} x_{id} y_{jd} \right)^2 \qquad (4.6)$$

subject to no constraint on X and Y.

$\left(m_{ij} - \sum_{d=1}^{k} x_{id} y_{id} \right)^2$ defined as squared error within observed and

predicted value. By using the gradient descent method unknown variables such as $x_{id} y_{id}$ must be learned to reduce the objective function. So with respect to independent variables $x_{id} y_{id}$, we find out the partial derivative of J.

$$\frac{\partial J}{\partial x_{iq}} = \sum_{j:(i,j)\in D} \left(m_{ij} - \sum_{d=1}^{k} x_{is} \cdot y_{js} \right) \left(-y_{jq} \right) \forall i \in \{1......a\}, q \in \{1......k\}$$

$$= \sum_{j:(i,j)\in D} \left(e_{ij} \right) \left(-y_{jq} \right) \forall i \in \{1......a\}, q \in \{1......k\}$$

$$\frac{\partial J}{\partial y_{jq}} = \sum_{i:(i,j)\in D} \left(m_{ij} - \sum_{d=1}^{k} x_{id} \cdot y_{jd} \right) \left(-x_{iq} \right) \forall j \{1......b\}, q \{1......k\}$$

$$= \sum_{i:(i,j)\in D} \left(e_{ij} \right) \left(-x_{iq} \right) \forall j \in \{1......b\}, q \in \{1......k\} \qquad (4.7)$$

ALGORITHM 4.1:   GD (RATING MATRIX: M, LEARNING RATE A)

Begin

1. Random initialization of matrices X and Y;
2. D = {(i, j):mij is observed};
3. while not (convergence)
4. do
5. Calculate every error $e_{ij} \in D$ as the perceived elements of M - XYT;
6. for every client-component pair (i,q) do

$$x_{iq}^{+} \Leftarrow x_{iq} + \alpha \cdot \sum_{j:(i,j) \in D} e_{ij} \cdot y_{jq}$$

7. for every product-component pair (j,q) do

$$y_{jq}^{+} \Leftarrow y_{jq} + \alpha \cdot \sum_{i:(i,j) \in D} e_{ij} \cdot x_{iq}$$

8. for every client-component pair (i,q) do $x_{iq} \Leftarrow x_{iq}^{+}$
9. for every product-component pair (j,q) do $y_{jq} \Leftarrow y_{jq}^{+}$
10. Check convergence condition:

end while
End Algo

In the above algorithm depicted in Algorithm 4.1, α is the step size which is greater than 0 and its value can be chosen using standard numerical method. Sometimes, its value set to small values. The execution is a convergence iteration process. This is known as gradient descent. In the above algorithm, to verify whether all the updates in the independent variable are occurred simultaneously, we used $x_{iq}^{+} y_{jq}^{+}$. Updating can be done by utilizing a sparse matrix by finding an error matrix $E = R - UV^{T}$ that contains the unperceived elements whose values are set to 0 (Charu 2016). It is a sparse matrix.

$$X \Leftarrow X + \alpha EY$$

$$Y \Leftarrow Y + \alpha E^{T} X \tag{4.8}$$

are the formulae to compute those updates. During the updations, we should take care of whether the entries are updated simultaneously or not according to the algorithm given above.

4.4.2 Single Value Decomposition

Single value decomposition can be defined as a form of matrix factorization in which the columns of matrix X and Y are constrained to be mutually orthogonal. The advantages of mutually orthogonal are:

1. Concepts can be completely autonomous to each other.
2. It is possible for them to be interpreted geometrically through a scatterplot.

Latent vector includes both positive and negative values, so semantic interpretation is generally difficult. It is generally simple to perform SVD with the utilization of eigenvalue strategies, for a fully specified matrix. Truncated SVD of rank k min {a, b} can be calculated as below:

$$M \approx Q_k \sum_k P_k \qquad (4.9)$$

which is used to factorize the rating matrix M.

The size of matrices Q_k, Σ_k, and P_k are a×k, k×k, and b×k, respectively. The k-largest eigenvectors of MM^T and M^TM are contained in the matrices Q_k and P_k respectively and the (non-negative) square roots of the k-largest eigenvalues of either matrix are contained in matrix Σk. The formula for the SVD can be formulated as:

$$\text{Minimize } J = \frac{1}{2} \| M - XY^T \|^2 \qquad (4.10)$$

Subject to: columns of X are mutually orthogonal, columns of Y are mutually orthogonal.

The presence of orthogonality constraints makes it different from the unconstrained factorization.

4.4.3 Non-Negative Matrix Factorization

The rating matrix that contains non-negative values, non-negative matrix factorization can be used here. It gives elevated level of interpretability in understanding the client–product interactions which is its major advantage. The factor X and Y always positive that makes this factorization method different from other methods.

The optimization formula for the non-negative can be formulated as:

$$\text{Minimize } J = \frac{1}{2}||M - XY||^{TT} \quad (4.11)$$

subject to: $X \geq 0, Y \geq 0$

Its most noteworthy interpretability favourable circumstances emerge in cases in which clients have a component to determine a preference for a thing, yet no system to indicate a dislike which is a non-negative appraisals matrix (Cichocki and Zdunek 2007; Hu et al. 2008, Choudhury, Mohanty, and Jagadev 2021).

4.5 Understanding the Matrix Family

In our previous discussions, we explain some different types of the matrix factorization method. The entirety of the previously mentioned optimizing definitions limit the Frobenius standards of the residual matrix (M - XY^T), taking into account to different imperatives on the rating matrix X and Y. The main focus of the objective function is to make XY^T inexact the rating matrix M as intently as could be expected under the circumstances. The imperatives on the factor matrix accomplish diverse interpretability properties. Truth be told, if we utilize different objective functions or constraint on the extensive group of matrix factorization, we can estimate better approximation values. It can be formulated as given below:

$$\text{Optimize } J = \left[\text{Objective function quantifying matching between M and XY}^T \right]$$

$$(4.12)$$

Subject to: Constraints on X and Y

Sometimes loss function can be defined as the minimization form of the objective function of a matrix factorization strategy. The aim of the objective function is consistently to constrain M to coordinate XY^T as intently as could be expected under the circumstances whereas the optimization detailing might be either a maximization or a minimization problem. The maximum-likelihood objective functions are the common maximization formulation used by some probabilistic matrix factorization methods and an instance of a minimization objective is Frobenius norm. In most cases, to forestall overfitting regularizers are added to the target capacity (Charu 2016).

In Table 4.1, we have given a rundown of different factorization models and their attributes/characteristics. By and large, the nature of the fundamental arrangement on the perceived elements can be reduced by adding conditions like non-negativity since it lessens the space of plausible arrangements. This is why unconstrained and optimal margin factorization is supposed to have the highest global optima. We use a constrained method to find the global optimum because it has better performance than under an unconstrained method. In addition to the impacts of overfitting, the exactness for

TABLE 4.1 Family of Matrix Factorization Methods

METHODS	CONSTRAINTS/ CONDITIONS	OBJECTIVES	PROS/CONS
PLSA	Non-negativity	Regularizers + Maximum Likelihood	Best for implicit feedback High semantic interpretability Lower over fitting in some situations Probabilistic interpretation Good quality solution Loses interpretability with both like/ dislike appraisals
Unconstrained	No constraints	Frobenius + regularizers	Over fitting Highest grade solutions Poor interpretability Regularization prevents Good for most matrices
NMF	Non-negativity	Frobenius + regularizers	Lower over fitting in some situations. Highest semantic interpretability Suitable for implicit feedback system. Suffer interpretability with both positive and negative ratings Good quality solutions
Max Margin	No constraint	Margin Regularizers + Hinge Loss	Good for discrete ratings Resists overfitting Highest quality solution Poor interpretability Similar to unconstrained
SVD	Orthogonal Basis	Frobenius + regularizers	Good option for dense matrix Good visual interpretability Out-of-specimen recommendations Poor in sparse matrix Poor semantic interpretability

the observed entries might be not quite the same as for unobserved elements. Sometimes, accuracy can be improved by non-negativity constraints on the unobserved entries in a certain field. The matrix contains negative entries; on that matrix we cannot use the factorization method as in non-negative matrix factorization. There is no single arrangement that can accomplish every one of these objectives. A cautious comprehension of the difficult area is significant for picking the right model (DeCoste 2006; Rennie and Srebro 2005).

4.6 Integrating Factorization and Neighbourhood Model

Neighbourhood model is a helpful system since it makes ready for a combination of neighbourhood model with other enhancement models, for example, the latent factor model can be integrated with the neighbourhood model in order to better optimization results (Charu 2016).

Let's take a ratings matrix M in which the global mean μ of the rating matrix has already been subtracted from all the entries and this matrix is known as a mean-centered matrix. Every expectation will be evaluated on mean-centered values and in a post-processing phase, these predicted values can be added with the global mean μ. We will return to the different parts of the model by making this supposition on the rating matrix M = [m_{ij}].

4.6.1 Baseline Estimator: A Personalized Bias-Centric Model

A personalized base-centric model predicts (mean-focused) appraisals in M simply as an expansion of client and thing predispositions. At the end of the day, evaluations are clarified totally by client's liberality and the thing fame, as opposed to explicit and customized interests of clients in things. Let the bias variable for user i be c_i^{user} and the bias variable for item j be c_j^{item}. Hence the prognostication for the model can be written as:

$$\overline{m} = c_i^{user} + c_j^{item} \tag{4.13}$$

Let D = {(i,j):m_{ij} is observed} which contains observed ratings in the rating matrix.

Now, c_{user} i and c_{item} j could be dictated by defining an objective function by the error $e_{ij} = m_{ij} - \overline{m}_{ij}$ as given below:

$$\text{Minimize } J = \frac{1}{2} \sum_{(i,j)\in D} \left(m_{ij} - \bar{m}_{ij}\right)^2 + \frac{1}{2} \left(\sum_{k=1}^{a} \left(c_k^{user}\right)^2 + \sum_{j=1}^{b} \left(c_j^{item}\right)^2 \right)$$

$$(4.14)$$

By using formulae of stochastic gradient descent method this optimization problem can be solved as given below:

$$c_i^{user} \Leftarrow c_i^{user} + \alpha \left(e_{ij} - \lambda c_i^{user} \right)$$

$$c_j^{item} \Leftarrow c_j^{item} + \alpha \left(e_{ij} - \lambda c_j^{item} \right) \qquad (4.15)$$

Regardless of its non-personalized nature, a pure bias-centric model can regularly give sensible forecasts. This is particularly the situation when the measure of ratings information is restricted. We set C_{ij} to the predicted value of $\bar{m}_{ij}$ after solving for the values of c_{user} i and c_{item}. This estimation of C_{ij} is then rewarded as a constant of all through this area instead of as a variable. Thus, the initial step for integration of models arrangement is to decide the constant quantity of C_{ij} by unravelling the non-personalized model (Charu 2016). This non-personalized model can be seen as a pattern estimator in light of the fact that C_{ij} is a rough baseline estimate to the estimations of the ratings m_{ij}. We can create a new matrix by deducting the estimate of C_{ij} from every perceived entries m_{ij} that can frequently be assessed by the vast majority of the models talked about in earlier sections. This area gives an explicit case of how neighbourhood models might be balanced with the utilization of the baseline estimator.

4.6.2 Neighborhood Portion Model

Equation for the neighborhood-based prediction model can be written as follows:

$$\bar{m}_{ij} = c_i^{user} + c_j^{item} + \frac{\sum_{l \in Q_j(i)} w_{lj}^{item} \left(m_{il} - c_i^{user} - c_j^{item} \right)}{\sqrt{|Q_j(i)|}} \qquad (4.16)$$

Here c_{user} = the user bias
c_{item} = the item bias

w_{ij}^{item} = the item-item regression coefficient between item i and item j (Charu 2016)

$Q_j(i)$ = the subset of the K-nearest items to item j have been rated by user i.

We can also replace $c_i^{user} + c_l^{item}$ with the constant value of C_{ij}, then the formula can be written as:

$$\bar{m}_{ij} = c_i^{user} + c_j^{item} + \frac{\sum_{l \in Q_j(i)} w_{ij}^{item} \left(m_{il} - C_{il} \right)}{\sqrt{|Q_j(i)|}} \tag{4.17}$$

Here C_{il} is a constant and c_i^{user} and c_j^{item} are bias variables which need to be optimized by the Stochastic gradient-descent approach as we have done in the previous section. This can be formulated as:

$$c_i^{user} \Leftarrow c_i^{user} + \alpha \left(e_{ij} - \lambda c_i^{user} \right)$$

$$c_j^{item} \Leftarrow c_j^{item} + \alpha \left(e_{ij} - \lambda c_j^{item} \right)$$

$$w_{ij}^{item} \Leftarrow w_{ij}^{item} + \alpha_2 \left(\frac{e_{ij} \left(m_{il} - C_{il} \right)}{\sqrt{|Q_j(i)|}} - \lambda_2 \cdot w_{ij}^{item} \right) \forall l \in Q_j(i) \tag{4.18}$$

By introducing item-item implicit feedback variables, we can have an improved neighbourhood model. If a user I rated an item j with its many neighbouring items, at this is point it ought to affect the anticipated rating. The effect is equivalent to $\frac{\sum_{l \in Q_j(i)} c_{ij}}{\sqrt{|Q_j(i)|}}$, where c_{lj} is the feedback variable and we divide $\sqrt{|Q_j(i)|}$ to adjust the fluctuating degrees of sparsity in various client products blends. Now the feedback is written as shown below:

$$\bar{m}_{ij} = c_i^{user} + c_j^{item} + \frac{\sum_{l \in Q_j(i)} w_{ij}^{item} \left(m_{il} - C_{il} \right)}{\sqrt{|Q_j(i)|}} + \frac{\sum_{l \in Q_j(i)} c_{lj}}{\sqrt{|Q_j(i)|}} \tag{4.19}$$

To figure out the gradient and determine the stochastic gradient-descent steps, one can make a least-squares optimization model corresponding to the error $e_{ij} = m_{ij} - \bar{m}_{ij}$, which can be formulated as:

$$c_i^{user} \Leftarrow c_i^{user} + \alpha \left(e_{ij} - \lambda c_i^{user} \right)$$

$$c_j^{item} \Leftarrow c_j^{item} + \alpha \left(e_{ij} - \lambda c_i^{item} \right)$$

$$w_{lj}^{item} \Leftarrow w_{lj}^{item} + \alpha_2 \left(\frac{e_{ij} \left(r_{il} - C_{il} \right)}{\sqrt{|Q_j(i)|}} - \lambda_2 \cdot w_{lj}^{item} \right) \forall l \in Q_j(i)$$

$$c_{lj} \Leftarrow c_{lj} + \alpha_2 \left(\frac{e_{ij}}{\sqrt{|Q_j(i)|}} - \lambda_2 \cdot c_{lj} \right) \forall l \in Q_j(i) \qquad (4.20)$$

For instance, a retailer may make the certain rating matrix dependent on clients who have appraised, browsed or purchased a thing (Charu 2016). This speculation is generally clear to consolidate in our models by evolving the last term of Equation (4.19) $\frac{\sum_{l \in Q'_j(i)} c_{lj}}{\sqrt{|Q'_j(i)|}}$. Here $Q'_j(i)$ is the arrangement of nearest neighbours of client I (in view of express evaluations), who have given some implicit type of certain feedback for thing j. This change can likewise be applied to the latent factor part of the model, in spite of the fact that we will reliably work with the streamlined suspicion that the certain input network is gotten from the rating matrix (Charu 2016).

4.6.3 Latent Factor Portion Model

Equation for the latent factor model where we make the integration of implicit feedback with rating information to perform prediction is given as:

$$\bar{m}_{ij} = \sum_{d=1}^{k+2} \left(x_{id} + \sum_{h \in I_i} \frac{f_{hd}}{\sqrt{|I|_i}} \right) \cdot y_{jd} \qquad (4.21)$$

I_i = the set of items rated by user i. Matrix size = a × (k+2)

$F = [f_{hs}]$ contains the implicit feedback variables

1s is contained by the $(k + 2)$th column of X *and* the $(k + 1)$th column of Y. The last two columns of F are 0s. The component $\sum_{s=1}^{k+2} x_{is} y_{js}$ includes the bias terms as the two columns of the factor matrix from the last holds the bias variables.

4.6.4 *Integration of the Latent Factor Model and the Neighbourhood Portion Model*

The equation after integrating the neighbourhood model and the latent factor model will be as given below:

$$\bar{m}_{ij} = \frac{\sum_{l \in Q_j(i)} w_{lj}^{item} \cdot \left(m_{il} - C_{il}\right)}{\sqrt{|Q_j(i)|}} + \frac{\sum_{l \in Q_j(i)} c_{lj}}{\sqrt{|Q_j(i)|}} + \sum_{d=1}^{k+2} \left(x_{id} + \sum_{h \in I_i} \frac{f_{hd}}{\sqrt{|I|_i}} \right) \cdot y_{jd}$$

(4.22)

The first two terms have come from neighbourhood portions and the last term come from latent factor portions.

The equivalent optimization issue, which reduces the total squared error $e_{ij}^2 = m_{ij} - \bar{m}_{ij}$ on the elements in the perceived set D, is as given below (Charu 2016):

$$\min\ J = \frac{1}{2} \sum_{(i,j) \in D} \left(m_{ij} - \bar{m}_{ij} \right)^2 + \frac{\lambda}{2} \sum_{d=1}^{k+2} \left(\sum_{i=1}^{a} x_{id}^2 + \sum_{j=1}^{b} y_{jd}^2 + \sum_{j=1}^{b} f_{id}^2 \right) +$$
$$\frac{\lambda_2}{2} \sum_{j=1}^{b} \sum_{l \in U_i Q_j(i)} \left[\left(w_{lj}^{item} \right)^2 + c_{lj}^2 \right]$$

(4.23)

subject to:

$(k + 2)^{\text{th}}$ column of X contains only 1s

$(k + 1)^{\text{th}}$ column of Y contains only 1s

Last two columns of F contain only 0s

λ and λ_2 are regularization parameters.

4.6.5 Solving the Optimization Model

A gradient descent method is utilized among the various optimization models mentioned earlier to take care of the enhancement issue. For this situation, the enhancement models are fairly unpredictable on the grounds that they contain plenty of terms, along with an enormous quantity of factors. A partial derivative as for every advancement quantity is utilized to determine the update step. The derivation of the gradient descent steps (Charu 2016) are omitted by us, and essentially state them here regarding the error esteems $e_{ij} = m_{ij} - \bar{m}_{ij}$

This can be formulated as:

$$x_{iq} \Leftarrow u_{iq} + \alpha \left(e_{ij} \cdot y_{jq} - \lambda \cdot x_{iq} \right) \forall q \in \{1....k+2\}$$

$$y_{jq} \Leftarrow y_{jq} + \alpha \left(e_{ij} \cdot \left[x_{iq} + \sum_{b \in I_i} \frac{f_{bq}}{\sqrt{|I_i|}} \right] - \lambda \cdot v_{jq} \right) \forall q \in \{1....k+2\}$$

$$f_{bq} \Leftarrow f_{bq} + \alpha \left(\frac{e_{ij} \cdot y_{jq}}{\sqrt{|I_i|}} - \lambda f_{bq} \right) \forall q \in \{1....k+2\}$$

$$w_{lj}^{item} \Leftarrow w_{lj}^{item} + \alpha_2 \left(\frac{e_{ij} \cdot \left(m_{il} - C_{il} \right)}{\sqrt{|Q_j(i)|}} - \lambda_2 \cdot w_{lj}^{item} \right) \forall l \in Q_j(i)$$

$$c_{lj} \Leftarrow c_{lj} + \alpha_2 \left(\frac{e_{ij}}{\sqrt{|Q_j(i)|}} - \lambda_2 \cdot c_{lj} \right) \forall l \in Q_j(i) \quad (4.24)$$

Here we set the unsettled entries in fixed columns of X, Y, and F.

We can follow the $(k + 2)$-dimensional vectored form to write the first three updates. We repeatedly loop all the observed ratings by gradient-descent method. The step size for variables corresponding to the neighbourhood portion model is controlled by α_2 whereas the step size for variables of the latent factor portion model is controlled by α. As is mentioned in the conditions of the optimization model, the column of X, Y and F should not be updated. When we are performing tuning and training on the accuracy of held out entries, the regularization parameter can be chosen. To stay away from poor performance by the latent factor and neighbourhood model portion, it is especially essential to utilize diverse step sizes and regularization parameters.

4.6.6 Perception about Accuracy

It can be seen that the integration of models performed better as compared to discrete models because the integration models can learn the variations in the datasets and adapt them. This method is similar to the hybrid recommender systems for joining various types of models. Cross-validations or hold-out techniques can also be used for picking off the relatives' weights. Albeit, in this scenario the integrated model more superior that isolated model because in the integrated model the overfitting is blocked as the bias variables shared between two segments. The use of the prediction function of Equation 4.20 automates the task of choosing values for the variables which, in turn, indirectly adjusts the significance of each segment of the models. Consequently, this method of integration provides a higher level of accuracy. However, integration of models performs marginally better than that of single value decomposition++ and its performance relied on the dataset. Again the neighborhood model has a higher number of parameters to be optimized as compared to single value decomposition++. If the dataset is not large enough optimum pros can't be extracted from the neighborhood model and in the case of a smaller dataset overfitting occurs due to an increase in no of parameters. In this regard, the best alternative between asymmetric models pure single value decomposition models with biases, single value decomposition ++ and neighborhood integrated factorization, relies on the size of datasets used. For integrated models, large datasets are required to bypass the overfitting. For exceptionally small datasets and exceptionally large datasets, latent factor models and neighborhood integrated factorization models are the best choices respectively. In most cases, single value decomposition ++ performs finer than pure single value decomposition with bias.

4.6.7 Integration of Latent Factor Models with Arbitrary Models

By integrating latent factor models with other type of model, such as content-based methods, we can create a hybrid recommender system. Generally, thing profiles might be accessible as depictions of items. Correspondingly, clients may have unequivocally made profiles depicting their inclinations. Let's $\bar{B}_i^{user}$ denotes profile for user i and $\bar{B}_j^{item}$ denotes profile for item j. Similarly, $\bar{M}_i^{user}$ denotes perceived appraisals for client i and $\bar{M}_j^{item}$ represents perceived appraisals for products j. Then the formula for the prediction function can be written as:

$$\bar{m}_{ij} = \left[\left(X + YF\right)Y^T\right]_{ij} + \beta F\left(\bar{B}_i^{user}, \bar{B}_j^{item}, \bar{M}_i^{user}, \bar{M}_j^{item}\right) \quad (4.25)$$

The first term is the latent factor portion and the second term is that model with which we shall integrate our latent factor model. β is the balancing factor. The second term is a parameterized function of the user profile, item profile, user ratings, and item ratings. In an attempt to minimize the error of prediction of the above equation, we can maximize the parameters of this function combining with the latent factor models. F() is a linear regression function in the case of the integration of the neighbourhood and latent factor models. To a notable extent the cutting edge can be improved in recommender frameworks by recognizing new wellsprings of information, whose prescient force can be coordinated with latent factor model utilizing that previously mentioned system.

4.7 Conclusion

This section examines various models for model-based collaborative filtering. Apart from all these some advantages of a model-based collaborating system are space-efficiency, where the learned model size is a lot smaller than a unique appraisals matrix. Therefore, the space prerequisites are frequently very low and the other one is avoiding overfitting where the model-based strategies can frequently help by its outline approach. Moreover, regularization techniques can be employed to make strengthen these models. This method applies machine learning methods and statistic strategies for the mining rating matrix. In spite of the fact that this methodology does not give the result which is as exact as the mix of CF and CBF approaches, it can take care of the issue enormous databases and sparse matrix. Sometimes classifications issues are faced by a collaborative filtering method. So, we have to use some sort of generalization whenever we are working with classification models with the implementation of a collaborating filtering method. One outstanding exemption occurs when latent factor models solve this problem exceptionally. The utilization of various kinds of factorization matrix by this model, so as to anticipate appraisals. We use these factorization methods by considering the type of target values and the imperatives on their performance matrix. Furthermore, they may have diverse exchange offs regarding exactness and over-fitting. One of the best models among the collaborating filtering methods is the latent factor model. Considering the choice of objective functions and enhancement requirements, various types of latent factor models have been advanced by the researchers. We can integrate other models such as neighborhood models with the latent factor models in order to obtain the profits of both of the models, resulting in the generation of a more powerful model.

References

Agarwal, D., & Chen, B. C. (2009). *Regression-based latent factor models.* In *Proceedings of the 15th ACM SIGKDD International Conference on Knowledge Discovery and Data Mining*, 19–28.

Aggarwal, C. C. (2015). *Data Mining: The Textbook.* Springer.

Basilico, J., & Hofmann, T. (2004, July). *Unifying collaborative and content-based filtering.* In *Proceedings of the twenty-first international conference on Machine learning*, 9.

Charu, A. C. (2016). *Recommender System.* Springer.

Choudhury, S. S., Mohanty, S. N. & Jagadev, A. K. (2021). Multimodal trust based recommender system with machine learning approaches for movie recommendation. *International Journal of Information Technology.* https://doi.org/10.1007/s41870-020-00553-2

Cichocki, A., & Zdunek, R. (2007). *Regularized alternating least squares algorithms for non-negative matrix/tensor factorization.* In *International Symposium on Neural Networks Springer*, Berlin, Heidelberg, 793–802.

DeCoste, D. (2006). *Collaborative prediction using ensembles of maximum margin matrix factorizations.* In *Proceedings of the 23rd International Conference on Machine Learning*, 249–256.

Garanayak, M., Mohanty, S. N., Jagadev, A. K., & Sahoo, S. (2019). Recommender System Using Item Based Collaborative Filtering(CF) and K-Means. *International Journal of Knowledge-based and Intelligent Engineering Systems*, 23(2), 93–101.

Garanayak, M., Sahoo, S., Mohanty, S. N., & Jagadev, A. K. (2020). An Automated Recommender System for Educational Institute in India. *EAI Endorsed Transactions on Scalable Information System*, 24(2), 1–13. doi.org/10.4108/eai.13-7-2018.163155

Hofmann, T., & Puzicha, J. (1999). Latent class models for collaborative filtering. In IJCAI (Vol. 99, No. 1999).

Hu, Y., Koren, Y., & Volinsky, C. (2008). *Collaborative filtering for implicit feedback datasets.* In *2008 Eighth IEEE International Conference on Data Mining, IEEE*, 263–272.

Kuanr, M., & Mohanty, S. N. (2018). Location based Personalized Recommendation systems for the Tourists in India. *International Journal of Business Intelligence and Data Mining,* In Press.

Kuanr, M., Mohanty, S. N., Sahoo, S., Rath, B. K. (2018). Crop Recommender System for the Farmers using Mamdani fuzzy Inference Model. *International Journal of Engineering & Technology*, 7(4.15), 277–280.

Rennie, J. D., & Srebro, N. (2005). *Fast maximum margin matrix factorization for collaborative prediction.* In *Proceedings of the 22nd International Conference on Machine Learning*, 713–719.

Sarwar, B., Karypis, G., Konstan, J., & Riedl, J. (2001). *Item-based collaborative filtering recommendation algorithms.* In *Proceedings of the 10th International Conference on World Wide Web*, 285–295.

Shani, G., Heckerman, D., Brafman, R. I. & Boutilier, C. (2005). An MDP-based recommender system. *Journal of Machine Learning Research*, 6(9), 1265–1299.

Thi, D. M. P., Dung, V. N., & Loc, N. (2010). *Model-based Approach for Collaborative Filtering.* In *6th International Conference on Information Technology for Education (IT@EDU2010).*

5

RECOMMENDER SYSTEMS FOR THE SOCIAL NETWORKING CONTEXT FOR COLLABORATIVE FILTERING AND CONTENT-BASED APPROACHES

R.S.M. LAKSHMI PATIBANDLA, V. LAKSHMAN NARAYANA AND AREPALLI PEDA GOPI

Vignan's Nirula Institute of Technology & Science for Women, Guntur, India

B. TARAKESWARA RAO

Kallam Haranadhareddy Institute of Technology, Guntur, India

Contents

5.1 Introduction

Recommended Systems (RS) are a range of digital tools and associated techniques that permit the recommendation of those items that are of most probable interest to the user. The guidelines include different facets of decision-making, such as what to do, whether to listen to music or what to learn online. The general word "topic" applies to what is recommended to users by the program. The RS usually works on a certain class of products (e.g. DVDs and news) and thus, the architecture, the interactive user interface, and the principal approach of recommendations for the creation of a particular category of object, are intended to render useful and effective clues. RS deliberately targets users who lack the technical information or experience to recognize, for example, the potentially vast amount of alternatives provided by a platform. One of the most common examples of this might be a book selection program offering recommendations for consumers seeking to buy a novel. On the popular Amazon.com website, for example, the company employs an RS to personalize the online store. Since these tips are typically customized, a variety of specific ideas are open to various people or communities. However, the reviews are often not customized [1].

These are also cheaper to produce and are generally found in papers or articles, which give lists of, for example, the top ten novels, CDs, and others. Although RS research in certain cases may be helpful and effective, it typically fails to tackle such kinds of non-personalized proposals. Specific recommendations are given as collections of objects in their simplest form. When implementing this assessment, Recommender Systems try, based on consumers' expectations and defects , to determine which goods or services are the most appropriate. Recommender Systems offer consumers with details regarding their desires to fulfill such a computational function, conveyed directly, e.g. as product reviews or inferred by the user's interpretation of behavior [2–5]. For example, Recommender Systems may use browsing to a specific product page as a symbol of the inferred interest for the goods on that list. A very basic insight started the creation of advisory systems: individuals always depend on the opinions of other people as they make regular, everyday decisions. For example, when selecting a book to be read, it is normal to rely on the recommendations of friends and colleagues; when selecting a film to view, the employers are usually read and rely on film reviews made by a film critic who appears in the newspaper they are reading, using letters of recommendation.

The first RSS has implemented algorithms to use feedback created by the users group and to send them to "good" users or users searching for advice to reproduce this behavior. The ideas were appreciated by particular consumers or individuals of common preferences. This technique is known as mutual screening, because it assumes that if the active user has previously responded to other users, the other comments submitted by specific users will be similarly applicable to the active user [6]. As websites for e-commerce began to grow, recommendations resulting from searching the entire range of options increased. It is hard for users to make the right choice from the broad variety of websites available (products and services). The unprecedented increase in and abundance of web-based information as well as rapid advances in modern e-business platforms (sales of products, comparing products, auctions, etc.) has also overwhelmed users. The abundance of available alternatives began to threaten the well-being of customers rather than generating a profit.

5.2 Traditional and Latest Systems Recommender

A huge amount of obscure information has become available with the rapid expansion of the internet and the growing penetration of e-commerce networks. Data collection is no longer a problem; however, useful knowledge is meaningfully gathered and provided to the customer [8–11]. Recommendation mechanisms have been established to better bridge the void between knowledge selection and analysis, to sort out all accessible material, and to send the most relevant recommendations to consumers. The program is intended to boost the capacity and quality of this method. The greatest obstacle with this sort of program is to strike the ideal balance between the recommender and the person receiving the recommendation; that is, it creates and finds the compromise between their needs.

Recommender Systems are defined as information structures that arrange relevant details for a single person depending on their profile. A recommending program usually applies the user profile to a single reference role, attempting to forecast a user's view about a specific object that has not yet been considered. The present-day major driver promoting system usage is e-commerce websites, which use various techniques to find more viable products and increase their clients' sales volumes.

5.2.1 Collaborative Scanning and the Scanning of Content

Two main approaches to RS have been identified: collective and content-based filtering. The collective search recommendation is based on analyses of related users for favorite items [13]. With the use of content-based filters, artifacts similar to those the consumer has already encountered and checked will be evaluated.

5.2.2 Partnership Screening

This method of filtering is Recommender based on user expectations predictions that lead to encounters with other users. Such a filtering system generally offers the consumer more surprises for his positive advice and can include entirely meaningless details in certain situations. Because they interpret data more than any code function, collaborative filter systems aim to involve people in their filtering schemes [14].

The first introduction to this filtering system offers recommendations focused on products encountered by consumers with the same consumption history as the new consumer. This is found mostly in online trading networks such as Amazon and Stock Broker [15–17]. The problem with the recommendation approach can be overcome using the material in the shared filtering process ,where

the user receives only identical information products. This approach does not, however, answer certain issues such as adding new products to the framework which would be approved after reading and reviewing by several users [18, 19]. Another concern is the question of those customers who do not express interest with other group leaders. An individual customer would also have no colleagues on which to construct the mutual advice system.

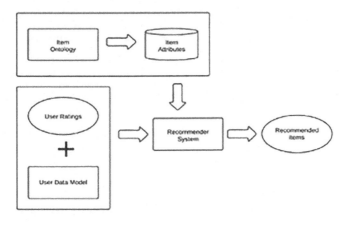

Figure 5.1   Content-based filtering

5.2.3 Material-Dependent Screening

This filtering strategy is grounded in the notion that the consumer needs to see items similar to those they have seen in the past. The customer-service relationship can be defined by knowledge about a particular consumer experience and data that can be related to that knowledge. This method utilizes content-based filtration techniques including extracting keywords and latent semantics (Figure 5.1).

For example, a computer-related course student program book can be recommended. This is the foundation of content-based recommendations that do not allow the use of the interaction between users to identify the information as opposed to collaborative filtering. For this reason, a content-based recommendation will not generally cause the consumer to be surprised, because the consumer will either explicitly or implicitly assume the connection and the means used by the method to recommend it.

5.2.4 Recommendation System Gaps

5.2.4.1 Dark Drop
With approved schemes, Cold Start issues are more popular. At the beginning of the program, the question occurs (there is no review from other users). Typically, the preferred framework contrasts the user profile with certain comparisons. The question is called gray sheep. In cases where an individual has unusual tastes, the advice will not be identical, because there are no "near environs."

The consumer of this profile is not readily associated with other members of the program in the shared filtering framework, rendering it difficult to selected products. If the consumer has an unusual profile, it is difficult to recommend products in the content-based filtering framework, since under the model recommendations are more common. For instance, whether or not the program understands the usage of science and oceanography, it will quickly prescribe such objects to the consumer, even though only unusual things have been assessed.

5.2.4.2 Late Determination Until individual reviews have been posted, a new item cannot recommend a new post. Throughout the collective filtering method, this issue is defined explicitly. When a product feature has not been implemented with a consumer test or recommendation, it cannot be recommended. Through the process of content-based filtering, the contents of an object are necessary to allow the consumer to create a recommendation.

5.2.4.3 Evaluations Sparse If only a few users are available and many items are to be evaluated, tests can be very small and users of this kind can be difficult to find. This problem is easily identified in collaborative filtering because filtering is completely based on the user's evaluation of the post. The policy does not depend on the number of users and items in content filters, but rather on their profiles and information.

5.2.4.4 High Specification Specialization It is advised that only products that are identical to those previously tested by the customer should be identified. Recent papers cannot be discussed. The recommendation is focused not on the original user profile in shared filtering, but instead on their behavior and interactions with other apps. That is because the advice needs suspense. This is rarely necessary to recommend items that are not related to the consumer profile. For content-based filtering, this question occurs because the Recommender material often belongs to the same user profile category. There are more likely to be surprises in collaborative filtering as similar users could evaluate completely different items to those seen by the original user.

When the number of users, objects, and tests is too high, any program that measures user interactions in real time may have a very long response time, demanding the use of scarce computing resources. This is a general issue in both directions. The question is more apparent in collective filtering, however, since both participants and all of the objects are used to measure. Calculations are rendered for one user only and associated artifacts in content-based filtering, taking all attributes into account.

5.2.5 Evaluation Methods in Recommender Systems

While identifying proposed programs as scientific analysis, it is important to consider the methodologies for program evaluation. A contrast of the predictions produced with the actual observations made by the consumer is the best way of assessing the Recommender Systems. This is accomplished by deleting a certain assessment, which is accompanied by a calculation of the excluded assessment by the preferred method and ultimately evaluating all values.

To test the efficiency of the proposed method before general commercial implementation, tests are necessary to validate the appropriateness of the forecasts to a specific use. The most commonly used criteria in the literature are given below for the assessment of the proposed schemes.

It is essential to remember that one particular recommendation framework might be best adapted to increase the dataset or market domain. Only by testing and evaluating the outcomes will it be possible to determine which of the guidelines are better included in a context.

The Root Mean Square (RMS) is the best way of calculating the discrepancy among an expected value and the real value in statistics. RMS calculates the medium square to calculate the meaning discrepancy (Table 5.1).

TABLE 5.1 MAE vs RMS

	ITEM 1	ITEM 2	ITEM 3
Current	3.0	5.0	4.0
Estimated	3.5	2.0	5.0
Difference	0.5	3.0	1.0
MAE	$= (0.5 + 3.0 + 1.0) / 3 = 1.5$		
RMS	$= \sqrt{((0.5^2 + 3.0^2 + 1.0^2) / 3)} = 1.8484$		

The RMS value for a set of parameters $\{x_1, x_2, \ldots, x_n\}$ are calculated as.

$$x_{rms} = \sqrt{\frac{x_1^2 + x_2^2 + \ldots + x_n^2}{n}}$$

The Mean Absolute Error (MAE) is calculated as.

$$MAE = \frac{1}{n}\sum_{i=1}^{n} f_i - y_i = \frac{1}{n}\sum_{i=1}^{n} e_i$$

5.2.6 Precision, Recall, and Fall-Out

Precision tests the quality of the recommendations provided on the Recommender programs and is calculated by the collection of all the Recommender goods by the amount of recommended objects currently deemed valuable to the consumer. The system's accuracy indicates exactly how similar the prediction is to the user's real assessment (Figure 5.2).

Recall shows the number of interesting items to the user that appear in the recommendation list in comparison to all relevant items (Figure 5.3).

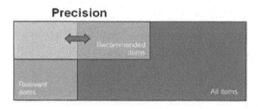

Figure 5.2   Precision

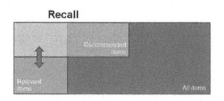

Figure 5.3   Recall

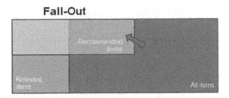

Figure 5.4 Fall-Out

Fall-Out is the proportion of non-relevant items recommended in comparison to all non-relevant items (Figure 5.4).

5.3 Educational Institutions and Social Networks

The program Recommender is known as a certain algorithm that can forecast this or that knowledge based on student data [8–10].

Four phases are included in the data-processing process:

- Collection of data – the student-based course management scheme requires an automated reporting framework in general and all the records are processed in the database;
- Software pre-processing – washing and code transfer in the appropriate format;
- Apply data analysis algorithms – using this basic phase or special resources to promote and openly extend the data processing process;
- Interpretation and assessment of the findings – the application of the policymakers' decisions.

The e-learning application monitors the history of the operation of each individual. The operation background may provide authorization on the server, job execution, etc. Such observations can be included in the Recommender structures as data sources. The valuable features resulting from the use of data analysis techniques allow teachers to determine the actions, and to view them correctly.

The user data needed for the recommendation framework may be explicit and tacit. Based on the onsite questionnaire results, the obvious profile of the student is established. The tacit profile of a student consists of his/her website behavior such as browsing time, copying the text from the page, and others. In terms of their expertise and preferences, each student is special due to its characteristics.

For the proposed program we must give three instructions:

- Advice for a new course – whether a person with a specific topic needs to propose a new course;
- Assistance during passage – when the person needs to take the course dependent on success and involvement in the specific recommendation;
- Course changes – a recommendation to develop and modify the course based on explicit and tacit feedback obtained during the course from students.

A mixture of the conventional approach to online education and social networking such as Facebook is one of the preferred frameworks. All social recommendation mechanisms are typically based on either an algorithm for shared screening or a content-driven algorithm. Recommend the program to use the user's generated stuff, the user's bookmark, etc.

Supported solutions are often used for online content on social networks. The route material is the foundation for the recommendation framework. Attributes for a specified framework are the main units that are evaluated.

Virtual courses demand that students work and take part in the project individually. Only a few thousand students will join the course, but the amount of students is always the lowest.

No one can be questioned on their own, there is also an issue, as adjustments cannot be done easily and some of the content is delivered in a video format because they are planned. This may be asked if the content is challenging to understand for certain students.

A method that combines classical learning with elements of social interaction and a recommendation scheme that examines the material required for resolving the problems defines the actions of the participant of the course and the resulting topic of personal recommendations (Figure 5.5).

5.4 Model of the User of Scientific Recommender

In this case, the definition of long-term and short-term patterns ("top" and "trend") can be created, in addition to identifying the user's information need. Certain applications can include top information units

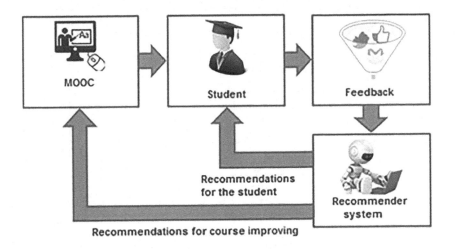

Figure 5.5   A place for recommender systems in remote education

most commonly checked for or accessed in the long term (quarter and semester). In the short run, individuals famous within a couple of days and weeks can be included. Based on the sphere of human activity, including details, the preview window can be modified more precisely.

This chapter offers the use of a joint approach for making recommendations to increase their pertinence, incorporating the above facts into consideration. The Recommender tuple duration must take into account human peculiarities, deciding the number of objects to be handled concurrently with 7±2. The following parts of the motorcade shall be (Figure 5.6):

1. S got, like the present, by users;
2. The IU, resulting from the most frequent;
3. Longer-term ("tops") pattern IU;
4. S linked to the current developments in the short term.

Items (1) and (2) are information needs that are retrievable and are expected to contribute the most to the desired collection (60–70%). Items (3) and (4) shall consist of data requires and occupy 30–40% of the tuple number. The final diagram of the double structure is shown in Figure 5.7.

The design of a program of research or schooling may take into account their specific characteristics. The operator of these structures

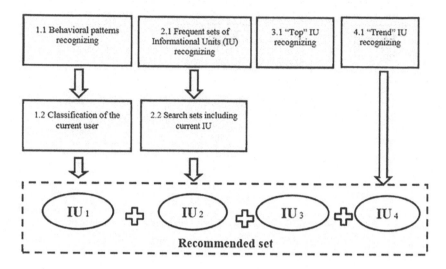

Figure 5.6 Construction of a pertinent set algorithm

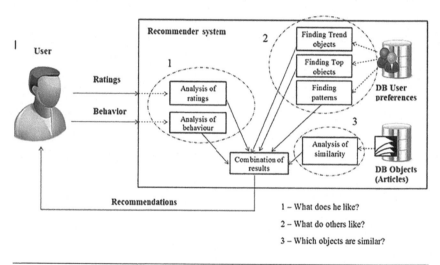

Figure 5.7 The scheme of user interaction with the scientific advisory system

communicates with the knowledge units and the outcomes of the analysis, as can be seen from the explanations of the above science networks.

The key characteristics of the individual will be age (date of birth), academic history, degree, and title (if available) in science guidance systems. The company's membership (name) and its position (country + town) are also visible.

The scientific outcome may contain numerous keywords that relate to a specific UPD (e.g. the paper, the monograph, the patent, and the doctoral thesis), the ISDN number, or the ISBN, that can be published on a given date in a particular journal (magazine, version, edition)).

The consumer engages with the findings or the characteristics of the study, conducting different behavior styles (creating, processing, awarding tests, uploading, etc.). Any action (where and when) is a space-time feature.

In addition to the behavior in the program, the individual can even be the creator or co-author of a different kind of arrangement to decide the experimental outcome. This model is generally meant to be viewed in a seminal network (graph), which enables heterogeneous connections. The entity (or entities), the research findings, the operations, and the organizations that are intertwined are then illustrated and are entirely intertwined.

A diagram that is completely connected should be depicted in a semantic network, i.e. the subject domain knowledge model, which has the shape of a graph guided, the vertices of which correspond to the domain artifacts, and the arcs (edges) define the relationships between them. According to training details, users can use a reference index, i.c. the representation of a semantic network, to create decisions using the framework to limit the number of connections representing the number of knowledge units (scientific results).

Figure 5.7 is a graph of the user-scientific advisory system.

The feedback framework utilizes two databases – the user's favorite database and the database for research tests.

Personal details that you include in the questionnaire on the platform can be viewed as the principal data sources for the consumer and operation log. This is automatically reported by the program based on what the consumer has achieved in the scientific method. The user's behavior in the scientific method can be split into two distinct categories: research and publishing.

There are several measures to receive a review from the end customer. The database for the study would be the user's records, the registration details, and the activities taken in the program. A consumer community of common preferences based on a profile provides recommendations. Defining such groups tends to define the user's expectations, i.e., to make them more important. This

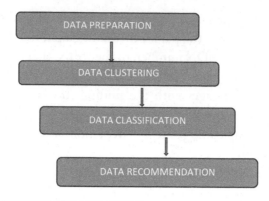

Figure 5.8   The sequence of actions to issue recommendations

process is referred to as clustering. A current person would be allocated the account to the highest one by classifying. In this way, it is recommended that the consumer understands which class is the student and whether other participants have common interests (Figure 5.8).

The analysis of published scientific works (NR) is a key value that need to be gathered and which can be dependent on further study. This table displays the outcomes of the registered user's research activities. As there are 20 works written in the sum of contents, the majority of researchers in the present reality are happy. The written material's names and keywords need to be handled and stored in separate areas to complete the description more accurately. The keywords listed by the author of the research are also taken into account.

The person reading the science research will consider important principles in the chart. Thirty plays are taken each month as a quantity of readable literature. The title and keywords of the reading material are also stored and preserved for classification in separate fields. The keywords listed by the author of this research will also be taken into account (Table 5.2).

The authorized levels of the users with different parameters are considered. The parameters are considered as numbers for all values (Table 5.3). The illustration is given in Figures 5.9 and 5.10.

The total number of attributes in the user preference database was 1279 in our experiment, with 9 attributes describing a person, 520 attributes describing their scientific activity, and 750 attributes describing their activity in the system.

TABLE 5.2　Information on the Authorized User

USER FIELD	COMMENT	TYPE
ID_user		Number
Degree		Number
Rank		Number
Age		Number
Education		Number
R_Works	Relation with workplace	Number
Organization		Number
City		Number
Region		Number

TABLE 5.3　Information on Scientific Activity of the Person

SCIENTIFIC ACTIVITYFIELD	COMMENT	TYPE
R_author=publishes	Publishes 20 articles per year	Number
ID_SC_Result		Number
Title	5 words	Text
Annotation		Text
Keywords	10 keywords	Text
Type		Number
UDC		Number
Date		Number
Refers to	5 keywords	Number
HP_organization		Text

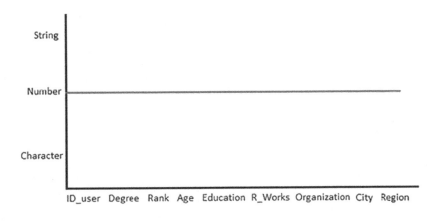

Figure 5.9　Authorized user parameters

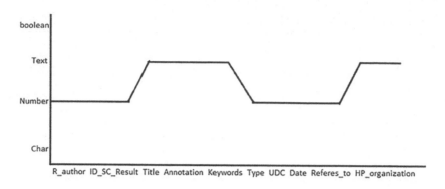

Figure 5.10   Scientific levels

5.5 Conclusions

This work has culminated in the knowledge model for users of science systems. The models are described in two ways: one as a semantic network (the whole one in connection type) and the reduced one. The interaction model is a simulation of a semantic network, taking into account the shortcomings of the transmission volumes which define the number of knowledge units used by the device user (scientific results). The developed interface model would also be used in addressing the issue of automated categorization and the creation of guidelines for users of science systems.

References

[1] K. Haruna, M. A. Ismail, A. B. Bichi, S. Danjuma, H. A. Kakudi, and T. Herawan, "*An overview on the state-of-the-art comparison of the three-contextualization paradigms,*" in *The 8th International Conference on Information Technology (ICIT 2017)*, May 17–18 (Amman, Jordan, 2017).

[2] F. Ricci, L. Rokach, D. Shapira, and P. B. Kantor, *Recommender Systems Handbook* (Springer Science+Business Media, LLC, 2011).

[3] P. A. Klemenkov, *Informatics, and Applications* 7(3), 14–21 (2013).

[4] A. Tejeda-Lorente, C. Porcel, E. Peis, R. Sanz, and E. Herrera-Viedma, *Information Sciences* 261, 52–69 (2014).

[5] See supplementary material at http://www.netflixprize.com/community/viewtopic.php?id=1537 for see discussion about the Netflix Prize and dataset.

[6] O. G. Ivanova, Y. J. Gromov, V. E. Didrih, and D. V. Polyakov, *Bulletin of Voronezh Institute of the Federal penitentiary service of Russia* 2, 49–54 (2011).

[7] D. E. Palchunov, and E. A. Uljanova, *Novosibirsk State University Journal of Information Technologies* 8(3), 6–12 (2010).

[8] D. Jannach, M. Zanker, A. Felfernig, and G. Friedrich, *Recommender Systems An Introduction* (Cambridge University Press,USA, 2010).

[9] C. Romero, S. Ventura, and E. Garcia, *Computers and Education* 51(1), 368–384 (2008).

[10] S. Fazeli, B. Loni, H. Drachsler, and P. Sloep, "*Which recommender system can best fit social learning platforms?*" in *Proceedings of the 9th European Conference on Technology Enhanced Learning, Graz, Austria, 2014* (LNCS, Springer Verlag, 2014), Vol. 8719, pp. 84–97.

[11] V. Del Castillo Carrero, I. Hernan-Losada, and E. Martin, "*Prototype of content-based recommender system in an educational social network,*" in *Proceedings of the 1st Workshop on Learning through Video Creation and Sharing at EC-TEL 2014*, Graz, Austria (2014).

[12] See supplementary material at http://helpiks.org/2-99885.html for see a model of semantic net's.

[13] A. Eckhardt, "*Various aspects of user preference learning and recommender systems,*" in *Proceedings of the Dateso 2009 Workshop, Spindleruv Mlyn, 2009*, edited by K. Richta, J. Pokorny, V. Snasel (Dateso 2009, Spindleruv Mlyn, Czech Republic, 2009), pp. 56–67.

[14] RSM Lakshmi Patibandla, "Regularization of graphs: Sentiment classification," in Mohanty, S. N., Chatterjee, J. M., Jain, S., Elngar, A. A. and Gupta, P. (eds.) *Recommender System with Machine Learning and Artificial Intelligence: Practical Tools and Applications in Medical, Agricultural and Other Industries* (pp. 373–386), 2020 Scrivener Publishing LLC.

[15] Choudhury, S.S., Mohanty, S.N., and Jagadev, A.K. Multimodal trust based recommender system with machine learning approaches for movie recommendation. *International Journal of Information Technology* (2021). https://doi.org/10.1007/s41870-020-00553-2.

[16] Garanayak, M., Sahoo, S., Mohanty, S. N., and Jagadev, A. K. An Automated Recommender System for Educational Institute in India. *EAI Endorsed Transactions on Scalable Information System*, 24(2), 1–13. doi.org/10.4108/eai.13-7-2018.163155.

[17] Garanayak, M., Mohanty, S. N., Jagadev, A. K., and Sahoo, S. Recommender System Using Item Based Collaborative Filtering(CF) and K-Means, *International Journal of Knowledge-based and Intelligent Engineering Systems*, 23(2), 93–101 (2019).

[18] Kuanr, M., and Mohanty, S. N. Location based Personalized Recommendation systems for the Tourists in India, *International Journal of Business Intelligence and Data Mining* (2018) Press.

[19] Kunar, M., Mohanty, S. N., Sahoo, S., and Rath, B. K. Crop Recommender System for the Farmers using Mamdani fuzzy Inference Model. *International Journal of Engineering & Technology*, 7(4.15), 277–280 (2018).

6

RECOMMENDATION SYSTEM FOR RISK ASSESSMENT IN REQUIREMENTS ENGINEERING OF SOFTWARE WITH TROPOS GOAL–RISK MODEL

G. RAMESH

GRIET, Hyderabad, India

P. DILEEP KUMAR REDDY

S. V. College of Engineering, Tirupati, India

J. SOMASEKAR

Gopalan College of Engineering and Management, Bangalore, India

S. ANUSHA

KSN Government Degree College for Women, Ananthapuramu, India

Contents

6.1 Introduction

Requirements engineering plays a crucial role in the software development life cycle of any project. During this phase, the requirements of customers or clients are gathered in order to build high-quality software that meets their intended requirements. When there are issues with the requirements, it leads to the unsatisfactory solution and deterioration of customer satisfaction. It also can lead to a failure in software projects. Therefore, in the requirements engineering phase, it is inevitable to have validations and risk analysis. It is very challenging to have an automated approach to have risk analysis. However, the literature found useful contributions towards it. In [1] goal-based risk analysis is studied based on the Tropos goal risk model. Other goal-based risk analysis models found in the literature are KAOS, i*, GBRAM, and Tropos. Of these, the Tropos model is found to be suitable and widely used.

The solution provided in [1] has limitations as it does not consider Satisfaction (SAT) and Denial (DEN) variables to have probabilistic or quantitative values. This limitation is overcome in this chapter by providing numeric values associated with the values of DEN and SAT variables. We proposed an algorithm to achieve risk analysis in the requirements engineering phase. Loan Origination Process (LOP) is the case study related to banking software taken for experiments in this chapter. This case study provides goals, events, and treatments suitable for our work. They are based on the three-layered approach, which contains layers such as goals, risks, and treatments. Our contributions are as follows.

- We proposed and implemented a methodology based on the Tropos Goal Risk model with a probabilistic approach to mitigate risks in candidate solutions.
- An algorithm is proposed to achieve probabilistic risk analysis in the requirements engineering phase using the LOP dataset.
- A prototype application is built to show the efficiency of the proposed model which caters to the needs of software engineers who are in a systems analyst role to have thorough investigation on the risks of requirements gathered. It demonstrates the process of risk mitigation and producing best optimized solutions.

The rest of the chapter is set out as follows. Sections 2 and 3 provide a review of the literature and the problem formulation respectively. Section 4 presents the proposed methodology. Section 5 presents the implementation details and experimental results. Finally, section 6 provides some conclusions and offers thoughts on the possible directions of future research.

6.2 Related Work

This section reviews literature on the goal risk models and risk analysis in the requirement engineering phase. The concept of risk priority using the goal risk model is explored in [2, 3]. They studied the Tropos goal model in order to enhance it. Probabilistic goals and obstacles is the approach followed in [4] for assessing risks associated with software requirements. In [5] the requirements engineering phase is explored with security requirements according to the standards provided by ISO 27001. A similar kind of study is made in [6] with the concept of ontology. The Tropos methodology for software engineering is studied in [7]. It is the basis for Tropos goal-based risk analysis and it is also close to the work of this paper.

The need for requirements engineering for adaptive systems is investigated in [8] for the better optimizations of the runtime of a system. The partial goal satisfaction and the reasons for it is investigated in [9] as part of requirements engineering. Risk assessment in requirements engineering with a goal-based approach is found in [1]. The Tropos methodology is explored with security context in mind, as studied in [10]. Model Driven Engineering (MDE) is employed in [11] for risk analysis. Security requirements engineering [12], risk and goal methods [13], the impact of requirements engineering phase in agile software development models [14], different methods or models for requirements engineering [15] and challenges in the requirement validations [16] are among the important aspects found in the literature. Apart from this, in [17], the software engineering process is scrutinized to identify potential risks. Safety and risk analysis in requirements engineering is explored in [18, 19], respectively, while the [20] focuses on goal-oriented risk analysis. The review of the existing literature revealed good contributions towards risk analysis in requirements engineering. In [1], the goal-based risk analysis needs to be improved with qualitative analysis. This is the reason why this paper extends goal based risk model in order to minimize risks further.

6.3 Problem Formulation

In software engineering, it is important to validate requirements. During the requirements engineering phase, it is essential to have some formal way of identifying risk and eliminating it early on so as to avoid propagating it to the next phase of the life cycle. Goal-oriented risk analysis provides consideration goals of stakeholders and associated events and treatments to have a better analysis of risk. The requirements gathered as part of the software project may have hidden risks. Checking the risks manually may take longer time and it may also lead to manual mistakes. An automated approach can reduce time and helps to ensure the more accurate assessment of risks. As presented in [1], there are many models for goal-oriented risk analysis. These include Tropos, KAOS, GBRAM and i*. Of these models, Tropos is used widely and the problem with existing models employing it is that they used qualitative values to DEN and SAT variables. This can be extended further with the adoption of a probabilistic or quantitative approach to have further mitigation of risks in the requirements engineering phase. This is the optimization problem considered in this book chapter.

6.4 Proposed Goal-Based Risk Analysis

The methodology provided in this section is influenced by the original Tropos goal risk model, which has adopted a layered approach in risk analysis. The aim of the methodology is to guide software engineers to analyze risk in the requirements engineering phase to help stakeholders to make well-informed decisions. The goal risk model presented in Figure 6.1 is used in making a risk analysis. The SAT and DEN variables representing satisfaction and denial are considered to have both qualitative and quantitative values. In this instance, N means not satisfied, F means fully satisfied and P means partially satisfied.

In this book chapter we have considered a quantitative or probabilistic approach and provided numeric values associated with the P, which indicates partial satisfaction or denial. This has resulted in producing more optimized solutions that lead to further cost reductions. It is based on the three-layered approach of the Tropos model. The layers include goals, risks and treatments.

The layer at the top is called the goal layer or asset layer. The goals are provided by the stakeholders of project. In the case of LOP, bank management has provided the goals. In the same fashion, events and treatments are provided by them. Events are the responses of

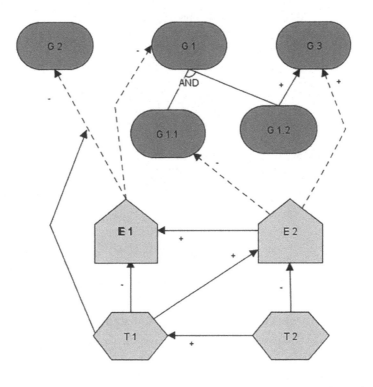

Figure 6.1   Layered approach for goal-based risk analysis

customers that may have an associated risk. For instance, the customer may provide fake identity certificates that lead to potential
risks to the banker. Each of such events needs certain reactions,
known as treatments. Each treatment has associated costs and risks.
The goals can be combined with operations such as AND/OR in
order to have subgoals (see Figure 6.2). Each event has its likelihood and severity. The likelihood attribute may have different values
like remote, occasional, frequent, reasonably probable and extremely
unlikely. Severity indicates the degree of affecting goals. It has four
levels as shown in Table 6.1.

Every event is nothing but an incident that may have an associated
potential to cause loss or harm. An event can be expressed with two
variables, known as SAT and DEN. These are otherwise known as
impacts of events, and this impact may be either positive or negative.
The impact values are presented in Table 6.2.

TABLE 6.1   The Four Severity Levels of Events

S NO.	LEVEL	MEANING
1	+	Stronger
2	++	Highly Stronger
3	-	Weaker
4	--	Highly Weaker

TABLE 6.2   Shows Impact of Events

S. NO.	IMPACT	MEANING
1	+	Positive impact
2	++	Highly positive impact
3	-	Negative impact
4	--	Highly negative impact

As mentioned earlier, the SAT and DEN variables can be associated with the impact values. If ++ is associated with SAT, then it is interpreted as the event has high positive impact on the satisfaction. In the same fashion if ++ is associated with the DEN variable, it indicates that the event has high negative impact on the denial. The treatment layer has a set of treatments and the required countermeasures. The countermeasures are, however, associated with certain cost or risk. They can be decomposed with AND/OR operators. A treatment may be able to reduce risk but it incurs cost.

As presented in Figure 6.2, which is adapted from [1], the LOP dataset used in the banking software project is modelled with goal-risk orientation. The three-layered approach is followed for better risk analysis. The model is represented as {N, R, I} tuple. N and R denote set of nodes and the relationships among the nodes, respectively, while I denotes impact relations. As the goals are used by stakeholders to build software, they are subject to risk analysis before going on to the next phase of the development. Every goal may lead to Satisfaction (SAT) or Denial (DEN). SAT and DEN can have qualitative values such as Full (F), Partial (P) and None (N). These values should be such that F>P>N. In this chapter we introduced a quantitative or probabilistic approach towards the P value of SAT and DEN variables. It is expressed with values between 0 and 1. The value 1 refers to fully satisfied or denial while the value zero refers to the fact that

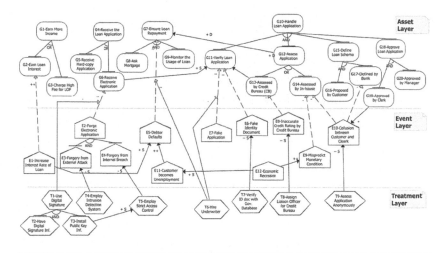

Figure 6.2   Goal–Risk model for LOP module of banking software

there is no satisfaction or denial. The intermediate values between 0 and 1 (exclusive of 0 and 1) show the degree or probability of satisfaction or denial.

6.4.1 Probablistic Risk Analysis (PRA) Algorithm

The proposed algorithm is known as Probabilistic Risk Analysis. It needs a goal risk model (as shown in Figure 6.2) and the degree of acceptable risk as inputs. After qualitative and quantitative analysis, the algorithm produces best-optimized solutions that can be considered to choose the best one that reduces risk and cost according to goal-based models. The LOP dataset is considered to evaluate the algorithm.

As presented here, the algorithm is able to consider both cost and risk. It has a set of candidate solutions. After analyzing cost and risk, it produces different combinations with associated cost and risk. The probabilistic approach enables to have more fine-grained solutions to come up. In other words, the risk is mitigated in various solutions and the final set of solutions that incur less cost and risk are presented to end users. This algorithm is implemented with a prototype application built in Java to demonstrate the utility of the proposed goal based risk analysis model.

ALGORITHM 6.1:   PROBABILISTIC RISK ANALYSIS

Algorithm: Probabilistic Risk Analysis
Inputs: Goal risk model m, risk r
Output: Solutions with less risk FS

1. cost[]=0
2. risk[]=0
3. solutions[]=0
4. solutions = computeCandidates(m)
5. cost = computeCosts(m)
6. risk=computeRisks(cost,m)
7. For each solution s in solutions
8. estimatedRisk=computeRisk(solution)
9. estimatedCost=computeCost(solution)
10. IF estimatedRisk is acceptable risk r THEN
11. add s to FS
12. END IF
13. End For
14. Return FS

6.4.2 Dataset Description

Loan Origination Process (LOP) is the dataset prepared based on a European project called SERENITY 2 [21]. The case study of LOP has many goals. The first goal is receiving a loan application and the final goal is loan approval. However, the goals can be provided in a different order so as to have permutations and combinations of solutions. The candidate solutions can be considered for risk analysis based on the goal risk model proposed in this chapter.

As shown in Figure 6.3, there are many goals, corresponding events and treatments. The goal model considered is based on the model presented in Figure 6.2. This dataset is subjected to probabilistic risk analysis to reduce risk and produce most optimized solutions.

6.5 Implementation and Results

A prototype application is built using a Java platform to show the usefulness of the proposed methodology. The application takes a LOP

G1 Earn_More_Income
G2 Earn_from_Loan_Interest
G3 Charge_High_Fee_for_LOP
G4 Receive_Loan_Application
G5 Receive_Hard_copy _App.
G6 Receive_Electronic_App.
G7 Ensure_Repayment_of_Loan
G8 Ask_Mortgage
G9 Monitor_Usage_of_Loan
G10 Handle_Loan_Application
G11 Verify_Loan_Application
G12 Assess_Application
G13 Assessed_by_Credit Bureau
G14 Assessed_by_In-house
G15 Define_Loan_Schema
G16 Proposed_by_Customer
G17 Defined_by_Bank
G18 Approve_Loan_Application
G19 Approved_by_Clerk
G20 Approved_by_Manager
E1 Increase_Interest_Rate_of_Loan
E02 Forge_Electronic_Application
E03 Forgery_from_External_Attack
E04 Forgery_from_Internal_Breach
E05 Cust_Fails_to_Fulfiill_Loan_Schedule
E06 Fake_Identity_Document
E07 Fake_Application
E08 Credit_Bureau_Ignorance
E09 Mispredict_Monetary_Cond.
E10 Collusion_Customer-Clerk
E11 Cust._Unemployment
E12 Economic_Crisis

T01 Use_Digital_Signature
T02 Have_Digital_Signature_Inf
T03 Install_Public_Key_Inf
T04 Employ_Intrusion Det. Sys.
T05 Employ_Strict Access Control
T06 Hire_Underwriter
T07 Verify_ID_doc_with_Gov_DB
T08 Assign_Liaison_Officer for CB
T09 Train_Internal_Actuary
T10_Assess_App._Anonymously

Figure 6.3   LOP dataset reflecting goals, events and treatments

dataset as input and performs cost and risk analysis using a goal-based risk analysis approach. It considered both qualitative and quantitative means of analyzing risks to have better optimizations in terms of solutions. The proposed model considers a layered approach with goals in the top layer, risks/events in the middle layer and treatments in the bottom layer.

As presented in Figure 6.4, the LOP dataset is loaded. LOP experiments are presented in this section though the application also supports other datasets. Once the dataset is loaded dynamically, a model is built with different layers according to the goal risk model presented in Figure 6.2.

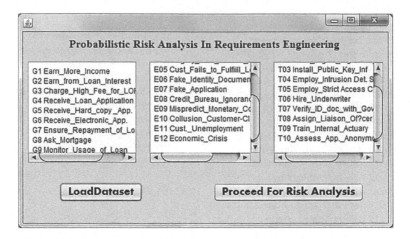

Figure 6.4 Prototype with UI

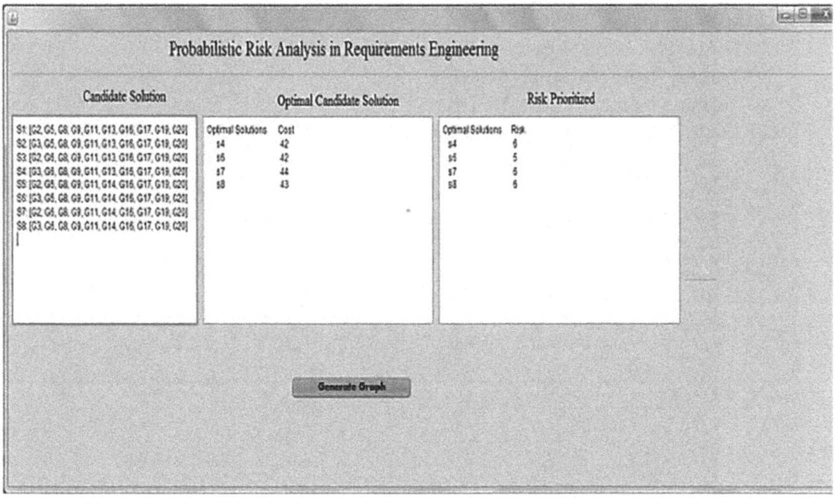

Figure 6.5 Probabilistic risk analysis producing optimal candidates and final solutions

The three-layered approach is employed to model and analyze the case study under consideration. Candidate solutions are generated in order to have formal analysis of risks and costs involved with each solution. Then the optimal solutions are identified with the proposed algorithm.

As presented in Figure 6.5, there are a total of 8 candidate solutions. Each solution is made up of a set of goals in a particular order. Each of these goals, in turn, is associated with certain events and treatments, as provided in Figure 6.2. Based on the analysis of risk and cost, final solutions are obtained that have less risk.

As presented in Figure 6.6, the treatments are considered to understand the cost and risk dynamics. This has led to the knowledge to identify the best-optimized solutions.

As can be seen from Figure 6.7, it is evident that the solutions are provided with associated cost and risk. The solution denoted as S6+C5

Treatment	Cost	S4				S6	S7/S8			
		C1	C2	C3	C4	C5	C6	C7	C8	C9
T02	2	S	N	N	S	N	S	N	N	S
T03	2	S	N	S	S	N	S	N	S	S
T04	1	N	S	S	S	N	N	S	S	S
T05	2	N	S	N	S	N	N	S	S	S
T06	4	S	S	S	S	S	S	S	S	S
T07	3	S	S	S	S	S	S	S	S	S
T08	2	S	S	S	S	N	N	N	N	N
T09	3	N	N	N	N	S	S	S	S	S
T10	2	N	N	N	N	S	S	S	S	S
Total Cost		13	12	12	16	12	16	15	15	19

Figure 6.6 Reflecting both qualitative and quantitative analysis

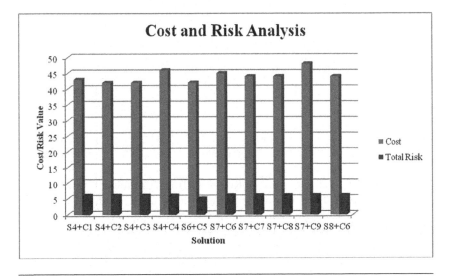

Figure 6.7 Reflects different solutions with associated cost and risk

has revealed minimum cost and risk. Therefore, it is considered to be the most optimal solution of those available. In this fashion, the goal-oriented risk analysis with the proposed methodology is able to reduce risk by considering the best solutions.

6.6 Conclusion and Future Work

Goal-based risk analysis in requirements engineering helps in identifying potential risks in software requirements. The early detection of risks can provide a better means of eliminating them and also avoids the unnecessary investment of time and effort. It also leads to the production of high-quality software. In this chapter, we proposed a goal-based risk analysis model with probabilities related to the values of SAT (Satisfaction) and DEN (Denial) variables. Our model extends the Tropos goal risk model, which is aimed at analyzing cost and risk in the requirements engineering phase. Most of the existing solutions considered only qualitative risk analysis with the probabilistic approach missing. In this chapter we proposed an algorithm for probabilistic risk analysis that contains both qualitative and quantitative risk analysis in order to achieve the better optimization of solutions. The LOP dataset of bank software is used for experiments. We built a prototype application to demonstrate the proof of the concept. The empirical results from our model revealed the utility of the prototype in finding risks in requirements and optimizing solutions that can lead to customer satisfaction. In future, the methodology provided in this chapter can be extended to the risk analysis of the requirements of Software Product Line (SPL).

References

[1] Yudistira Asnar, Paolo Giorgini and John Mylopoulos. (2011). Goal-driven risk assessment in requirements engineering. *Requirements. Engineering.* 16, 101–116.
[2] K. Venkatesh Sharma and P.V. Kumar (2013). An efficient risk analysis based risk priority in requirement engineering using modified goal risk model. *International Journal of Computer Applications.* 73 (14), 15–25.
[3] K. Venkatesh Sharma and P.V. Kumar. (2013). A method to risk analysis in requirement engineering using Tropos Goal Model with optimized candidate solutions. *International Journal of Computer Science Issues.* 10 (6), 250–259.

[4] Antoine Cailliau and Axel van Lamsweerde. (2013). Assessing requirements-related risks through probabilistic goals and obstacles. *Springer*, 1–19.

[5] Kristian Beckers, Stephan Fabender, Maritta Heisel, Jan-Christoph Küster and Holger Schmidt. (2009). Supporting the development and documentation of ISO 27001 information security management systems through security requirements engineering approaches, 1–8.

[6] Amina Souag, Camille Salinesi, Isabelle Wattiau and Haralambos Mouratidis. (2013). *Using security and domain ontologies for security requirements analysis. The 8th IEEE International Workshop on Security*, 1–8.

[7] Mirko Morandini, Fabiano Dalpiaz, Cu Duy Nguyen and Alberto Siena. (2011). The Tropos Software Engineering *Methodology*, 1–31.

[8] Mirko Morandini, Loris Penserini, Anna Perini and Alessandro Marchetto. (2012). Engineering Requirements for Adaptive systems. *Requirements Engineering*, 1–28 .

[9] Emmanuel Letier and Axel van Lamsweerde. (2004). Reasoning about partial goal satisfaction for requirements and design engineering. *ACM*, 29, 1–11.

[10] Haralambos Mouratidis, Paolo Giorgini, Gordon Manson and Ian Philp. (2002). A Natural Extension of Tropos Methodology for Modelling Security, 1–13.

[11] Denisse Muñnante, Vanea Chiprianov, Laurent Gallon and Philippe Aniort. (2016). A review of security requirements engineering methods with respect to risk analysis and model-driven engineering. *International Cross-Domain Conference and Workshop on Availability.* 8708, 1–17.

[12] Ilham Maskani, Jaouad Boutahar and El Houssaini. (2016). Analysis of security requirements engineering: towards a comprehensive approach. *International Journal of Advanced Computer Science and Applications.* 7, 38–45.

[13] Shankarnayak Bhukya and Suresh Pabboju. (2016). Requirement engineering – Analysis of risk with goal method. *International Journal of Computer Applications.* 153, 37–42.

[14] Sehrish Alam, S. Asim Ali Shah, Shahid Nazir Bhatti and Amr Mohsen Jadi. (2017). Impact and challenges of requirement engineering in agile methodologies: A systematic review. *International Journal of Advanced Computer Science and Applications.* 8, 411–420.

[15] Mona Batra and Archana Bhatnagar. (2017). A comparative study of requirements engineering process model. *International Journal of Advanced Research in Computer Science.* 8, 740–745.

[16] Sourour Maalema and Nacereddine Zarour. (2016). Challenge of validation in requirements engineering. *Journal of Innovation in Digital Ecosystems.* 3, 16–19.

[17] Manju Geogy and Andhe Dharani. (2016). A scrutiny of the software requirement engineering process. *Global Colloquium in Recent Advancement and Effectual Researches in Engineering, Science and Technology.* 25, 406–409.

[18] Shaojun Lia and Suo Duoa. (2014). Safety analysis of software requirements: Model and process. *3rd International Symposium on Aircraft
Airworthiness*. 80, 153–164.

[19] K. Appukkutty, Hany H. Ammar and Katerina Goseva Popstajanova.
(2005). Software requirement risk assessment using UML. *IEEE*. 112,
1–4.

[20] Antoine Cailliau and Axel van Lamsweerde. (2012). A probabilistic
framework for goal-oriented risk analysis. *20th IEEE International
Requirements Engineering Conference (RE)*, Chicago, IL, 201–210.

[21] www.serenity-project.org

A COMPREHENSIVE OVERVIEW TO THE RECOMMENDER SYSTEM

Approaches, Algorithms and Challenges

R. BHUVANYA AND M. KAVITHA

Vel Tech Rangarajan Dr. Sagunthala R&D Institute of Science and Technology,
Chennai, India

Contents

7.1 Section I – Introduction

Recommender Systems are one of the most valuable applications of recent Machine Learning (ML). Amazon attributes over 35% of their revenue to recommendations, and companies such as YouTube and Netflix rely on them entirely to allow their users to discover new content. If the site is recommending things like articles, music, movies, people, or webpages, the underlying systems are based on recommender systems, which predict ratings or preferences [19] that a user might give to an item. Accordingly, such a system is recommending products and services to people based on both their past behavior and that of people with similar preferences.

There is a major difference between the search engine and the recommendation engine. A search engine is a program which searches and finds the items in a database based on keywords and characters as specified by the user, whereas the latter is a software tool which generates recommendations to the user by analyzing their previous preferences. Yet in practice the distinction is a little more blurred; the modern search engine tends to be personalized since it does not retrieve the information as found.

In the modern era, the Recommender System (RS) is a leading domain in the research field. It was introduced in the year of the mid-1990s to select a product from the abundant set of choices which, in turn, relates to the notion of 'Big Data'. Today's web is flooded with abundant information. Those users who have a lack of experience in purchasing a product may take a poor decision if the recommendation is not available on the site. After the launch of the recommender system in an e-commerce site, users are purchasing the product happily by looking at the additional choices or accessories for the product. Hence the customer satisfaction will be achieved through the recommendation engine.

One of the best examples of the recommender system under the e-commerce site is Amazon.in, if the user tries to purchase an item on the site then the user will be notified with a list of widgets as a recommendation by considering their past interaction. Also the recommendation information can be blended from like-minded users. Yet such systems are not only restricted to e-commerce sites and nor do they necessarily have to be physical objects. Recommender systems might also consider content [11], music, and videos. In today's trend, service-based recommendations can also be provided through a consideration of the service-based systems [5]. For instance, the familiar site Netflix will recommend new movies and TV shows to the user by considering both their past preferences and the preferences of other people. Even in dating apps, a recommendation is simply a person. A social net-working site such as Facebook provides a recommendation in the form of 'people you may know' by relating the mutual friends of a particular user. The recommender system can also be applied in educational sectors [32] for offering personalized course recommendation [44].

Badrul Sarwar [47] and Schafer [48] explains the applications and algorithms of recommender systems in an e-commerce application. Besides all these, the recommendations can be afforded to a group of people, in a system known as Group Recommendation System (GRS) [6]. D. Wu [8] and K. I. B. Ghauth [31] implemented the rec-ommender system in an e-learning environment. In [8], the recom-mendation is achieved by fuzzy tree matching-based hybrid learning activity and in [31], the vector space model is incorporated.

Hence the recommender system finds a relationship between user and item based on actions. Usually, there is no human intervention at all. The system recognizes that, statistically, people who buy genera-tors also buy starter fluid, hence it can use those historical patterns to recommend new items before they even know they want it. However, the discovery of patterns alone is not sufficient for personalized rec-ommendation. The critical step is extracting good quality and useful aggregate usage profile from the pattern. Hence Mobasher, B., Dai, H., Luo, T. et al. [45] incorporated the clustering techniques based on user transactions and page views that can be effectively used for real time web personalization.

When designing a recommender system, it is important to think about what sources of data we have about the user's interest, because even the best recommender system cannot produce good results unless it has an abundant amount of good data to work with. A recommender system starts with some sort of data about every user which is used to figure out an individual's tastes and interests. It will then merge the collected data with the collective behavior of similar users to recommend the stuff that the user likes. Chen, Chi-Hua [22] introduced the delicacy recommender system using effective decision support system and cloud-based recommender system. Cloud based recommender system incorporates the multiple document summarization which summarizes the comments of various delicacies and provide suggestion based on the user interaction and historical experience.

7.1.1 Working Principles of the Recommender System

In general, Recommendation engines are data-filtering tools that make use of algorithms and data to recommend the most relevant items to a particular user. And the recommendation system works, based on the interaction with the user-item.

  i. User-Item Interactions: The information will be retrieved about the users by analyzing the user's profile and preferences.
  ii. Exploration of diverse content: The information will be retrieved about the product by considering the characteristics such as ratings given for the product, the number of purchases made, likes and dislikes given for the selected product.

7.1.2 Architecture of Recommender System

Figure 7.1 illustrates the model architecture of the recommender system published by Amazon in the year 2003. Initially, the candidate generation phase holds the item which is interested by the user then it consults with another data store known as 'Item Similarities' and combines few items which are similar to the user interest. For example, if person A has been interested in the movie *A Quiet Place* and based on everyone else's behavior, it is shown that the people who like *A Quiet Place* might also like the movie *The Silence*. Thus, based on user interest, person A may get a recommendation that includes *The Silence*. Then the score has to be assigned to each candidate based on the consideration of the similarities between their preferences.

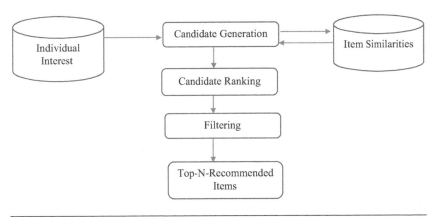

Figure 7.1 The architecture of the recommender system

The next phase is candidate ranking, where many candidates will appear more than once. The results of these two processes need to be combined. The ranking stage will access more information about the recommendation candidates, such as average review scores, which could be used to boost the results for highly rated or popular items. The filtering stage eliminates the recommendation for the items the user has already rated since the system should not recommend the things the user has already seen. Finally, the output of the filtering stage is then handed off to the display layer where a clear widget of product recommendation will be presented to the user. The candidate generation, candidate ranking, and filtering will reside inside the distributed recommendation web service.

7.1.3 Ways of Collecting Feedback

One way to gain knowledge of the customer is through explicit feedback. For example, asking the users to rate the recently purchased product on a scale of 1 to 5 stars or rating content in terms of 'thumbs up' or 'thumbs down'. In these cases, the system will ask for explicit feedback. But the problem with explicit ratings or feedback is that it always requires extra work from users; also, the data will be sparse. When the data are too sparse, this leads to a low-quality recommendation. Another way to understand the user's taste is through implicit

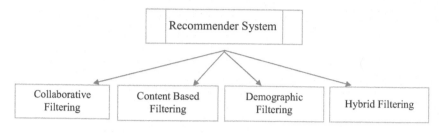

Figure 7.2   Classification of the recommender system

behavior, as suggested by Yang et al. [7]. This is a way of predicting the user by looking at the things they view and click; the recommender system will interpret it as an indication of a user's interest or disinterest. Compared to the explicit feedback, 'click' data, meaning that there is no issue of data sparsity to be dealt with. However, clicks are not always a reliable indication of interest.

7.1.4 Evaluating a Recommender System

A good recommender system makes effective use of both relevant and useful recommendations. The evaluation of the recommender system can be done in three ways: offline, online, and user studies. Initially, the performance of the recommendation engine has been evaluated and graded by considering its prediction capability. In some cases, an accurate prediction is still crucial while deploying a good recommendation engine. This chapter further explores the ways of evaluating a recommender system in the following section.

7.1.5 Classification of the Recommender System

The Recommender Systems can be classified in terms of Content-Based, Collaborative, Hybrid, and Demographic filtering System which will be briefly described in the upcoming sections. Figure 7.2 illustrates the classification of the recommender system.

7.1.6 Contribution of the Chapter

The main contribution of this work are as follows:

- Working Principle and architecture of the Recommender System

- Evaluation metrics of the Recommender System
- Exploration of methodologies adopted in the Recommender System
- Challenges identified in the Recommender system
- Comparative Study of Machine Learning Algorithm in Recommender System

7.1.7 Organization of the Chapter

This chapter is organized as follows. Section II narrates the variety of filtering techniques and the evaluation metrics applied in previous works for recommending the products elaborately, Section III illustrates the challenges identified in the recommender system. Finally, Section IV encompasses the comparative study of algorithms in the recommender system followed by the conclusion.

7.2 Section II – Related Works

Because of the increased usage of recommender systems in all the domains, there are plenty of recommendation algorithms. Thus, the recommendation algorithms need to be evaluated [9] based on offline and online to know the performance. People can also compare the revenue of e-commerce sites with and without incorporating the recommendation engine.

Rajani Chulyadyo [8] et al. explained the evaluation metrics where the offline evaluation is a broadly used technique that calculates the accuracy without involving the actual users. And it will be carried out based on splitting the data into training and testing sets. Ratings for the item can be predicted through the training set. The training set is bigger as it may have 80 to 90 percent of the data. Once the data are trained, the recommendation engine will be asked to make predictions about how a new user might rate some item that they have never seen before. To improve the performance of offline evaluation, instead of a single train and test split, create many randomly assigned training sets named k-fold cross-validation. Here individual training set or fold is used to train the recommender system independently and then the accuracy of the resulting system has to be measured against the test set. User ratings are predicted accurately by considering each fold's score and finally averaging them together. Though it takes a lot more computing power, the advantage is it will not end up in over-fitting. Also, if it is carried out for a small number of people it may come with the results of, how good the recommender system is at recommending things that the people already watched and rated. As mentioned above, however, the system has to recommend new things to people that they have not seen but which they might find interesting. However, that is fundamentally impossible to test offline. So, researchers cannot just test out new algorithms on real people or Netflix or Amazon.

On the other hand, in the online evaluation method, the user interacts with a recommender system and receives recommendations. And customer satisfaction with quality recommendations will be evaluated based on how users give feedback and how do they behave while interacting with the system. It can be illustrated as 'gathering data through user click'. But evaluations based on the online search is a little expensive and time-consuming. Herlocker, Jon, Konstan, Joseph [41] et.al introduced empirical results on accuracy metrics that provide initial insight into how results from different evaluation metrics might vary.

Montaner et al. [42] proposed a 'profile discovering evaluation procedure with collaboration'. The idea behind the technique is similar to profile discovery; in addition to that, however, it considers the opinions and recommendations of other users. It mainly incorporates the procedure of Profile Generation, Contact List Generation, Recommendation, and Profile Adaptation process. The initial step of 'Profile Generation' follows the profile discover procedure, obtaining as many items as evaluation from real-time users. The second step of the Contact List Generation generates the list of contacts by referring to the friends of the particular person. The third step of the recommendation process will recommend new items to the user through an analysis of their profile. In addition to that, the recommendation process can be done collaboratively by considering the opinion of neighbors or friends of a particular user. Finally, the evaluation of the recommendation system can be done by assessing user choices. Hence the above procedure implies that the evaluation is done based on gathering real-time information and it does not consider the user's opinion about the product available on the e-commerce site.

Gunawardana et al. [17] discussed the prediction accuracy calculation in the recommender system. In general, the people's myth about the recommender system is, if the system provides more accurate predictions then it is considered good. Hence there exist various algorithms to detect prediction in the recommender system. Predicting the accuracy can be divided into three cases, measuring the accuracy of predicting the ratings, usage, and ranking of items available on the site. It is important to measure the prediction error by comparing the expected result with the system generated output.

The straightforward approach is Mean Absolute Error (MAE). The formula to calculate MAE is given below in Equation 7.1.

$$MAE = \frac{\sum_{i=1}^{n} |y_i - x_i|}{n} \qquad (7.1)$$

Assume that the dataset has n ratings in the test set that must be evaluated. Besides, if the user gives the rating then it is denoted as 'x' and the system predicted rating is denoted as 'y'. The absolute value of difference has to be calculated to measure the error for rating prediction. It is also defined in terms such as the difference between the predicted and the actual rating. To calculate MAE, sum those errors across all n ratings in the test and divide it by the value of 'n' (the number of ratings) to obtain the average or mean. So, MAE is exactly defined [36] in such a way that, mean of the absolute values of the individual prediction error in the test set. An example of calculating MAE is given below in Table 7.1. MAE value is indirectly proportional to accuracy. Thus, the minimum MAE value leads to better accuracy.

TABLE 7.1 System Predicted Rating vs Actual Rating with Error Calculation

S.NO	PREDICTED RATING	ACTUAL RATING	ERROR
1	5	3	2
2	4	1	3
3	5	4	1
4	1	1	0

MAE = 2+3+1+0/4 -and the calculated MAE value is 1.5

A slightly more advanced way to measure accuracy is Root Mean Square Error (RMSE). Here the system will generate the predicted ratings, for which the true ratings are known. The formula to calculate the RMSE is given below in Equation 7.2, where p_i and o_i denotes the predicted and actual rating.

$$RMSE = \sqrt{\frac{\sum_{i=1}^{n}\left(p_i - o_i\right)^2}{n}} \tag{7.2}$$

RMSE = 4+9+1+0/4 and the calculated RMSE value is 1.87.

Table 7.2 exemplifies the calculation of RMSE. In this case, the errors are squared before they are averaged; thus, this gives greater weight to large areas. In general, the RMSE values will always be equal or higher than MAE. If there exists a greater difference, the greater the variance in the sample will be. If the calculated values of RMSE are equal to MAE, then the errors are of the same magnitude. Both the values can range from 0 to ∞. Both the RMSE and the MAE can be normalized, being named, in turn, Normalized RMSE (NRMSE) and Normalized MAE (NMAE). In addition to the above, the properties of predictive accuracy measures are described by Alexander Dekhtyar [28].

In addition to the above metrics, two important things to be noted while doing recommendation is Precision and Recall. Bellogín [20] implements the precision-oriented techniques, such as precision, recall, and nDCG experimentally. A recall is defined as a model of finding all the similar items and recommends them to the user; on the other hand, precision measures the capability of delivering the relevant elements with the least amount of recommendations. To provide a recommendation, there exist various filtering algorithm which is discussed briefly.

TABLE 7.2   System Predicted Rating vs Actual Rating with Squared Error Calculation

S.NO	PREDICTED RATING	ACTUAL RATING	ERROR
1	5	3	4
2	4	1	9
3	5	4	1
4	1	1	0

As defined by Goldberg [52] in 1992, Collaborative Filtering (CF) [34] is the information-filtering system and the filtering will be more effective when humans are involved in the system so the concept was clearly understood by Resnick [51] and implemented CF, by considering the like-minded users. Users who often rate the items alike must be identified. From the phrase, it is identified that if any like-minded user gives positive reviews or ratings to the product then it will be recommended to the user. But Yang et al. [7] identified the issue in the scenario when the user provides no ratings to the item. At times, it may lead to a Cold Start problem, which is one of the major challenges in the recommender system. Also, the Cold Start problem occurs when the new item is launched on the site or when the new user enters the system. So, the recommendation system may find it difficult to gather information about the item and users. The best solution can thus be provided in terms of considering the implicit ratings such as gathering data through 'Click Analysis'. And Vellino [26] compared the implicit ratings between the research paper's recommendation and movie recommendation.

The Collaborative Filtering (CF) model can also be defined in terms of recommending the products based on observing the historical connections between the user and item available on the site. It can be represented in terms of matrix named as interaction matrix denoted as m × n, where m represents the row of items and n represents the column of users. As mentioned by S. Gong et al. [30], Collaborative Filtering is divided into memory-based CF and model-based CF. John S. Breese David Heckerman Carl Kadie [50] explored the memory, model-based CF and also compared the efficiency of algorithms. However Xue, Gui-Rong, Lin et.al [38] combined the memory and model based approaches which takes advantage of both the methods by introducing smoothing based method. As a result, there is increased efficiency in providing recommendation.

Memory-based CF uses the user-item dataset to create a prediction. This method finds a set of users, also known as neighbors. Once the neighbors are identified, the recommendation engine will combine the preferences taken from the neighbors and produce a prediction. But the memory-based CF also deals with the known difficulty of handling a large number of users quite referred to as scalability issues.

TABLE 7.3 User and Item Rating

S.NO	USER	ONE INDIAN GIRL	REVOLUTION TWENTY20	THE GIRL IN ROOM 105
1.	Alexa	5	4	4
2.	Siri	5	?	4

Memory-based CF is further classified into Item-Item CF and User-Item CF. In User-Item CF, like-minded users are identified through similar ratings for similar items and can predict the missing rating. User and item-based prediction is explained by Manos Papagelis [37].

For instance, say Alexa and Siri have given similar ratings to the product. Table 7.3 refers to the User and Item Rating.

Hence the next step is to predict the unobserved rating for the user 'Siri' based on the rating observed by 'Alexa' considering the User-Item relationship.

Steps Identified to predict the missing rating in User-Item:

1. Specify the target. For instance, the target user is 'Siri'.
2. The next step is to find a like-minded user by considering a similar rating for similar items.
3. Isolate the item's rating to be predicted.
4. Finally, predict the rating for all unobserved items. If the predicted ratings range above the threshold value, then the items will be recommended to the user.

In item-based CF, instead of finding similar users, the aim is to find similar items to those already rated by the user. So, consider the example mentioned in Table 7.4, which has a single user rating.

Steps Identified to predict the missing rating in Item-Item:

1. Specify the target. In the above example, the target user is 'Alexa'.
2. Instead of finding like-minded users, find similar items with similar ratings which the target user has already rated.
3. Finally, predict the ratings and if the predicted value meets the range above the threshold, then recommend the items to the user.

TABLE 7.4 Individual User Rating

S.NO	USER	ONE INDIAN GIRL	REVOLUTION TWENTY20	THE GIRL IN ROOM 105	FIVE POINT SOMEONE	2 STATES
1.	Alexa	5	4	4	?	?

To overcome the drawback of the Scalability issue covers under memory-based CF, a model-based recommendation approach is introduced. It works based on developing a model on the dataset which contains ratings. It is enough to extract a few information points from the dataset, and that can be built as a model to provide a recommendation, with no necessity of taking the complete set of data every time. In the model-based approach, similarities between the user and items are calculated and then the stored similarity value is used to predict ratings. This model can be built using similarities between items rather than between users. At times, the predictions may not be accurate, since the entire dataset is not taken into consideration. However, the prediction's quality depends on the way the model is built. Combining the technique of memory-based and model-based collaborative filtering will always yield good results. Yu Li [39] applies a hybrid collaborative filtering method, to improve recommendation accuracy for Multiple interests and Multiple-content recommendation. The CF, based on item and user (CF-IU), is framed by combining the CF based on the item (CF-I) and CF based on user (CF-U), whereas Huang qin-hua [33] enhanced the collaborative filtering by applying the fuzzy technique with multiple agents. Vellino, André, and David Zeber [35] applied the hybrid and multi-dimensional for the recommendation of journal articles in scientific digital library.

As referred by Ricci [23], Content-based filtering (CBF) is the most widely used recommendation class. The major principle behind the CBF is the user-modeling process, where the user's interest is extracted from the site by knowing the items with which the user interacted. Such items can be textual as mentioned by Paik [46], and it could be emails or webpages as defined by Seroussi [25]. The items can even be layout information or XML tags, as explained by Shin [49]. CBF is a technique that compares the attributes of the user profile to make a recommendation. Also, it can be defined as the recommendation that will be provided based on the description of the product. It includes the techniques to estimate the relevance of the item by considering the user's previous rating. To find the similarity between the items, Term Frequency-Inverse Document Frequency (TF-IDF) can be adopted.

TF-IDF is a subdomain in Natural Language Processing (NLP). In this system, TF is defined as the frequency of the word in the current document to the total number of words in the document, whereas IDF is the total number of documents to the frequency occurrence of the document containing the word. Also, it allows a user-based personalization, hence it can provide the best choice of recommendation for each user. Here each item will be analyzed to extract features and, based on the extracted features, user models will be built and similarity calculation will be performed. The drawback with the CBF is that it requires more computing power. And the weakness in the CBF is it may lead to overspecialization so there is a probability of recommending the item that the user knows already. Here the sources of data are considered as 'Sentences'. To illustrate the TF-IDF calculation [54], consider the example given below,

Sentence 1= 'Recommender_system is a popular field of Data_Science and Recommender_system is efficient'

Sentence 2= 'Recommender_system has plenty of algorithms available and the algorithms are efficient '

Set of Words Collected by combining Sentence 1 and 2= {'plenty', 'and', 'of', 'efficient', 'available', 'is', 'algorithm', 'popular', 'has', 'the', 'field,', 'are', 'Recommender_system', 'a', ' ' , 'Data_Science'} here each word is mapped with the values of 1 to 16, respectively, including the whitespaces. Figure 7.3 illustrates the Set of Words formed by combining Sentence 1 and Sentence 2.

Figure 7.4 displays the frequency count of each word in the Sentence. '0' denotes the nil occurrence of a word and Figures 7.5 and 7.6 represent the Term Frequency Calculation and Inverse Document Frequency Calculation, respectively.

Figures 7.7 and 7.8 represent the graphical illustration of TF and TF*IDF, respectively. Content-based filtering is further classified into Pure CBF, CBF Separated, and CBF-combined as referred to by Beel [10]. In Pure CBF, the user model is compared with the recommendation candidates and CBF separated, works based on collecting

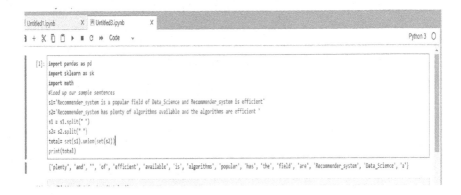

Figure 7.3 Set of words formed by combining Sentence 1 and Sentence 2

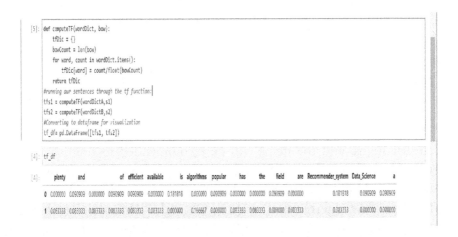

Figure 7.4 Frequency count of each word in the sentence

Figure 7.5 Term frequency calculation

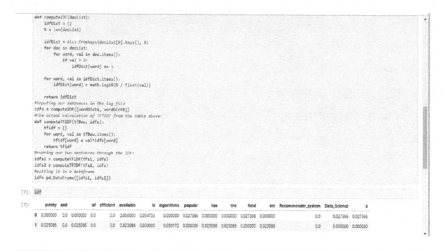

Figure 7.6   Inverse document frequency calculation

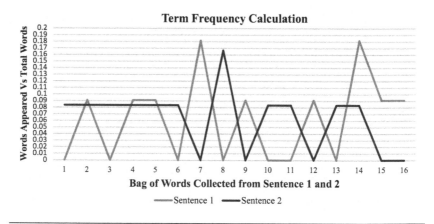

Figure 7.7   Graphical representation of term frequency calculation (TF-DF)

similar items, and the items will be recommended to the user. On the other hand, in the CBF-combined approach, the most similar user model will be recommended by considering the Pure CBF as its baseline. As suggested by Luis [27], CF and CBF can be combined in a system known as Hybrid Filtering [16] [27]. Since the Bayesian

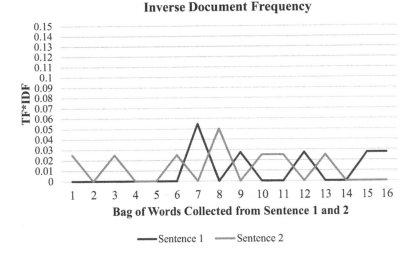

Figure 7.8 Graphical representation of inverse document frequency calculation (TF*IDF)

Networks have produced successful results in the domain of Artificial Intelligence (AI), the above-mentioned work combines the technique of CF and CBF along with Bayesian networks. By considering the data sparsity issue in the recommendation system this work has computed the ranking in realtime. And this method has proved successful in providing various recommendation tasks such as predicting missing ratings and finding relevant items.

In addition to Collaborative Filtering (CF) and Content-Based Filtering (CBF), the recommendation can also be provided through co-occurrence, as mentioned by Small [53] in 1973. Today's e-commerce sites of Amazon and Flipkart will always show the product recommendation with the familiar phrase 'Customers Who Bought This Item Also Bought....'. Here Amazon analyses the frequently purchased items by the user and provide choices to increase the revenue and customer satisfaction. Wang, Yi-Fan, Chuang, Yu-Liang et al. [40] implemented the personalized recommender system for the cosmetic business using content based, collaborative and data mining techniques.

Further, the approaches can be classified in terms of demography, in which the recommendation can be provided by considering the age, gender, and geographic location, but it is bit difficult in applying a real-time environment. The community-based filtering provides the recommendation by considering the users in the community who shares a common interest. Finally, the knowledge-based filtering is the intelligent one, which provides recommendation by analyzing the user needs.

7.3 Section III – Challenges of Recommender System

This section explores the various challenges arising in the Recommender System.

7.3.1 User Privacy

The demographic filtering recommends the items to the user by considering the age, gender, geographic location, etc. At times, the user may not be interested in revealing the details to the system and even if the details are collected from the user, then their privacy is negotiated. Demographic filtering and clustering may cluster the users with similar preferences hence the performance will be improved and the latency will be decreased [12].

7.3.2 Data Sparsity

In general, most users will not provide a rating to most of the items. Hence the data sparsity problem will occur in a collaborative filtering technique [3]. So, the construction of the user-item matrix will be a bit difficult in this case. Hence the solution to sparsity [21] can be solved by applying domain information.

7.3.3 Scalability

To provide a recommendation is challenging when the recommender system deals with huge datasets. Since some recommendation algorithms work extremely well with small datasets, but its efficiency may not work for large datasets.

7.3.4 Freshness

This problem occurs when the recommendation adopts the technique of global relevance. By considering the aspect of the most-liked item on the site the recommendation will be shown. But the downtime here is, the user may get the recommendation of the already seen product. As mentioned above, the recommendation is not intended to provide the already known items instead, to provide items of similar interest.

7.3.5 Cold-Start Problem

In general, the recommendation will be provided by considering the like-minded users and profile analysis. When a new item is introduced onto the site or when a new user enters, then the recommendation engine may find it difficult to provide a recommendation, called as Cold Start problem [14]. The solution for the problem can be provided by incorporating the global relevance where the system can retrieve the item based on its popularity.

7.3.6 Profile Injection Attack

Today's social media is flooded with various fake profiles. So, if any fake user enters the false rating to the item then the user may get an incorrect recommendation in which he/she might not be interested. It is also known as shilling attacks.

7.4 Section IV – Comparative Study of Algorithms in Recommender System

There are various algorithms available in Machine Learning to produce a recommendation and it needs to be analyzed [4] in terms of its benefits and detriments which would be helpful for the developers to identify the bottleneck in the algorithms. Table 7.5 portrays the comparative study of various algorithms with benefits and limitations.

TABLE 7.5 Comparative Study of Various Algorithms in the Recommender System

S.NO	ALGORITHM	DESCRIPTION	BENEFITS	DETRIMENTS
1.	K-nearest Neighbour (K-NN) [1]	K-NN falls under the collaborative filtering approach. It works by computing the similarity between the users and finds the K most like-minded user and then the recommendation will be provided.	K-NN algorithm is extremely simple to apply. It can be applied to both classification and regression. Analysis can be done through the historical data hence training steps need not be incorporated. It is also known as a lazy learning algorithm.	Inefficient in handling missing values. Though it is easy to implement, the performance of the algorithm will be deduced with a larger dataset.
2.	Matrix Factorisation [15, 29]	A popular approach for large scale, real world recommender is Matrix factorisation [24]. User-Item interaction matrix will be constructed based on the ratings given by the user to the items, and this data will be sparse. The matrix factorisation technique will reduce the dimensionality of the interaction matrix and construct square symmetric matrix n×n. Commonly used matrix factorisation methods are Singular Value Decomposition (SVD), Principal Component Analysis (PCA), and Probabilistic Matrix Factorisation (PMF).	Matrix factorisation uses implicit feedback if explicit feedback is not available. The recommendation is especially difficult when dealing with high-dimensional datasets, by applying the dimensionality reduction, the size of the datasets will be reduced.	SVD is difficult to interpret. Data loss may occur in PCA if the items are not properly selected. Data Standardization is mandatory before PCA.

(Continued)

TABLE 7.5 (CONTINUED) Comparative Study of Various Algorithms in the Recommender System

S.NO	ALGORITHM	DESCRIPTION	BENEFITS	DETRIMENTS
3.	Association Rules [18]	Association Rules are useful in recommending a product by connecting the edges of frequently purchased items. It is a rule-based machine learning method. The most known algorithms are Apriori algorithm and FP Growth.	Apriori algorithm is simple to implement and it doesn't require labelled data as it is completely unsupervised.	Apriori works well with similar datasets. FP Growth will consume more memory to process.
4.	Vector Space Model (VSM) [2]	The major difference between the Bag-of-words and Vector Space model is, Bag of words will retrieve the words, but it does not consider syntax and semantics and retrieves the information based on the Unigram approach. Whereas the Vector Space Model (VSM) is considered as the statistical model for information retrieval technique which is used to represent a document in terms of vectors containing feature and weight. The 'words' present in the document is referred to as feature, whereas 'value' is represented as weight. The weight is assigned with a binary value. So '1' indicates the term appeared in the document and '0' indicates the term is not present in the document. Also, the VSM model can have a frequency value, which denotes the number of times the term occurred in the document.	It is a Simple Statistical model. VSM can find the similarities of words across the document.	It may yield a 'false positive' result, by considering the substring of words, since the keyword to be searched should exactly match the document. There is a possibility of a 'false negative' result since the vocabulary words will not be taken into account.

(Continued)

TABLE 7.5 (CONTINUED) Comparative Study of Various Algorithms in the Recommender System

S.NO	ALGORITHM	DESCRIPTION	BENEFITS	DETRIMENTS
5.	TF-IDF [43]	Most useful algorithm for Natural Language Processing (NLP). It is a text vectorisation technique. As Machine Learning cannot deal with text, words can be converted into numbers by applying the Term Frequency and Inverse Document Frequency technique. TF calculates the frequency of the most frequent word. IDF is the total number of documents to the frequency occurrence of the document containing the word.	Major Applications are Information Retrieval and Keyword Extraction. Easy to compute the similarity between the documents.	Finds difficulty in capturing the semantics of the word. It doesn't consider the co-occurrence of another word in a document.
6.	Classifier [13]	A classifier is a technique that helps to input data to a specific category. It is the process of predicting the targets for given data points. Decision trees, Naïve Bayes, and ANN are the popular classifier algorithms.	Naïve Bayes classifier is easy to implement. It requires only a small amount of data for training. The decision tree is simple to implement whereas the Neural Network is comparatively faster in prediction.	Decision trees require much time to train the model. A Small change in the decision tree will affect the entire structure. Neural Network is expensive than traditional techniques.

7.5 Conclusion

This chapter has explored the usage of Recommender System in the modern era and how it has become unavoidable in all the domains. Users of social media have increased day by day. It is a way where we can find a larger number of users and their associated information. Hence the recommendation can be provided by analyzing the user profiles, the behavior of the user concerning the item, and so on. This chapter discusses how the recommendation algorithms need to be evaluated. It also narrates the various filtering technique in the recommender system. Collaborative filtering is efficient filtering in the recommender system. The CF is further divided into memory-based CF and model-based CF. When combining memory- and model-based CF, it provides a good and accurate recommendation. On the other hand, content-based filtering provides a recommendation to the user by constructing the user model. Major challenges in the recommender system are discussed and various algorithms are also elaborated as it needs to be analyzed with its merits and demerits. If the recommendation system is explained from the user's perspective, this provides exciting user experiences and satisfaction whereas, from the business perspective, revenue will be increased by displaying additional products for the customer.

References

[1] Gallego, A., Calvo-Zaragoza, J., and Rico-Juan, J. R. (2020), 'Insights into efficient k-nearest neighbor classification with convolutional neural codes', *IEEE Access*, Vol. 8, pp. 99312–99326.

[2] Chen, S., Chen, Y., Yuan, F. et al. (2020), 'Establishment of herbal prescription vector space model based on word co-occurrence', *The Journal of Supercomputing*, Vol. 76, pp. 3590–3601.

[3] Alhijawi, Bushra and Kilani, Yousef (2020) 'The recommender system: a survey', *International Journal of Advanced Intelligence Paradigms*, Vol. 15, No. 3, pp. 229–251.

[4] Dacrema MF, Cremonesi, P., Jannach, D. (2019), *'Are we really making much progress? A worrying analysis of recent neural recommendation approaches'*, In *Proceedings of the ACM Conference on Recommender Systems (RecSys)*, pp. 101–109.

[5] Xie, Fang et al. (2019), 'An integrated service recommendation approach for service-based system development', *Expert Systems With Applications*, Vol. 123, pp. 178–194.

[6] Fang, Guang, Lei, Su, Jiang, Di, and Wu, Liping (2018), 'Group recommendation systems based on external social-trust networks', *Wireless Communications and Mobile Computing*, Vol. 2018, Article ID 6709607.

[7] Longqi Yang, Yin Cui, Yuan Xuan, Chenyang Wang, Serge Belongie, and Deborah Estrin. 2018. *Unbiased offline recommender evaluation for missing-not-at-random implicit feedback.* In *Proceedings of the 12th ACM Conference on Recommender Systems (RecSys '18).* Association for Computing Machinery, New York, NY, 279–287.

[8] Wu, D., Lu, J., and Zhang, G. (2015), 'A fuzzy tree matching-based personalized E-learning recommender system', *IEEE Transactions on Fuzzy Systems*, Vol. 23(6), pp. 2412–2426.

[9] Kumar, Balraj and Sharma, Neeraj (2016), 'Approaches, issues and challenges in recommender systems: a systematic review', *Indian Journal of Science and Technology*, Vol. 9, pp. 47.

[10] Beel, Joeran, Gipp, Bela, Langer, Stefan, Breitinger, Corinna (2016), 'Research-paper recommender systems: a literature survey', *International Journal on Digital Libraries*, Vol. 17(4), pp. 305–338.

[11] Guan, Congying, Qin, Shengfeng, Ling, Wessie, and Ding, Guofu (2016), 'Apparel recommendation system evolution: an empirical review', *International Journal of Clothing Science and Technology*, Vol. 28(6), pp. 854–879.

[12] Khusro, S., Ali, Z., and Ullah, I. (2016), 'Recommender systems: issues, challenges, and research opportunities', In: Kim K., Joukov N. (eds) *Information Science and Applications (ICISA) 2016.* Lecture Notes in Electrical Engineering, Vol. 376. Springer.

[13] Ren, Y., Zhang, L. and Suganthan, P. N. (2016), 'Ensemble classification and regression-recent developments, applications and future directions [Review Article]', *IEEE Computational Intelligence Magazine*, Vol. 11(1), pp. 41–53.

[14] Zhao, Wayne, Li, Sui, He, Yulan, Chang, Edward, Wen, Ji-Rong, and Li, Xiaoming (2015), 'Connecting social media to E-commerce: cold-start product recommendation on microblogs', *IEEE Transactions on Knowledge and Data Engineering*, Vol. 28, pp. 1.

[15] Bokde, Dheeraj, Girase, Sheetal, Mukhopadhyay, Debajyoti. (2014). Role of Matrix Factorization Model in Collaborative Filtering Algorithm: A Survey. *International Journal of Advance Foundation and Research in Computer*, Vol. 1, pp. 2348–4853.

[16] Al-Hassan, Malak, Lu, Helen, and Lu, Jie. (2015), 'A Semantic Enhanced Hybrid Recommendation Approach: a Case Study of E-government Tourism Service Recommendation System', *Decision Support Systems*, Vol. 72, pp. 97–109.

[17] Gunawardana, Asela and Shani, Guy. (2015), 'Evaluating recommender systems', doi:10.1007/978-1-4899-7637-6_8.

[18] Solanki, S. K. and Patel, J. T. (2015), 'A Survey on Association Rule Mining', *2015 Fifth International Conference on Advanced Computing & Communication Technologies*, Haryana, 2015, pp. 212–216.

[19] Chulyadyo, Rajani & Leray, Philippe. (2014), 'A personalized recommender system from probabilistic relational model and users preferences', *Procedia Computer Science*, Vol. 35.

[20] Bellogín, A., Castells, P., Cantador, I. (2011) 'Precision-oriented evaluation of recommender systems: An algorithmic comparison'. In: *Proceedings of the ACM Conference on Recommender Systems (RecSys)*, pp. 333–336.

[21] Chen, Yibo and Wu, Chanle and Xie, Ming and Guo, Xiaojun. (2011), 'Solving the sparsity problem in recommender systems using association retrieval', *JCP*, Vol. 6, pp. 1896–1902.

[22] Chen, Chi-Hua and Yang, Ya-Ching and Shih, Pei-Liang and Lee, Fang-Yi and Lo, Chi-Chun. (2011), 'A cloud-based recommender system - a case study of delicacy recommendation', *Procedia Engineering*, Vol. 15, pp. 3174–3178.

[23] Ricci, F., Rokach, L., Shapira, B. and Kantor, P.B. (2011) *Recommender Systems Handbook* (Vol. I).

[24] Roberts, W.J.J. (2010), 'Application of a Gaussian, missing-data model to product recommendation', *IEEE Signal Processing Letters*, Vol. 17, no. 5, pp. 509–512.

[25] Seroussi, Y. (2010), 'Utilising user texts to improve recommendations', In De Bra, P., Kobsa, A., Chin, D. (eds) *User Modeling, Adaptation, and Personalization*. UMAP 2010. Lecture Notes in Computer Science, vol 6075. Springer, Berlin, Heidelberg.

[26] Vellino, Andre (2010). 'A comparison between usage-based and citation-based methods for recommending scholarly research articles', *Proceedings of the American Society for Information Science and Technology*, Vol. 47, pp. 1–2.

[27] de Campos, Luis, Fernández-Luna, Juan, Huete, Juan, and Rueda-Morales, Miguel. (2010), 'Combining content-based and collaborative recommendations: a hybrid approach based on Bayesian networks', *International Journal of Approximate Reasoning*, Vol. 51, pp. 785–799.

[28] Dekhtyar, Alexander (2009), 'CSC 466: knowledge discovery from data', Spring 2009.

[29] Koren, Y., Bell, R. and Volinsky, C. (2009), 'Matrix factorization techniques for recommender systems', *Computer*, Vol. 42(8), pp. 30–37.

[30] Gong, Song Jie et al. (2009), '*Combining memory-based and model-based collaborative filtering in recommender system*', *Pacific-Asia Conference on Circuits, Communications and Systems*, pp. 690–693.

[31] Ghauth, K. I. B. and Abdullah, N. A. (2009), '*Building an E-learning recommender system using vector space model and good learners average rating*', *2009 Ninth IEEE International Conference on Advanced Learning Technologies*, Riga, pp. 194–196.

[32] Drachsler, Hendrik, Hummel, Hans, and Koper, Rob. (2008), 'Personal recommender systems for learners in lifelong learning networks: the requirements, techniques, and model', *International Journal of Learning Technology*, Vol. 3, pp. 404–423.

[33] Huang, Qin-Hua, Ouyang Wei-Min (2007), 'Fuzzy collaborative filtering with multiple agents', *Journal of Shanghai University* (English Edition), Vol. 11, pp. 290–295.

[34] Schafer, J. et al (2007). 'Collaborative filtering recommender systems', *The Adaptive Web*, pp. 291–324.

[35] Vellino, André, and David Zeber (2007), '*A hybrid, multi-dimensional recommender for journal articles in a scientific digital library*', IEEE/WIC/ACM International Conferences on Web Intelligence and Intelligent Agent Technology-Workshops (2007), pp. 111–114.

[36] Adomavicius, G. and Tuzhilin, A. (2005), 'Toward the next generation of recommender systems: a survey of the state-of-the-art and possible extensions', in *IEEE Transactions on Knowledge and Data Engineering*, Vol. 17(6), pp. 734–749.

[37] Papagelis, Manos and Plexousakis, Dimitris (2005), 'Qualitative analysis of user-based and item-based prediction algorithms for recommendation agents, *Engineering Applications of Artificial Intelligence*, Vol. 18(7), pp. 781–789

[38] Xue, Gui-Rong, Lin, Chenxi, Yang, Qiang, Xi, WenSi, Zeng, Hua-Jun, Yu, Yong, and Zheng, Chen (2005), '*Scalable collaborative filtering using cluster-based smoothing*', In *Proceedings of the 28th Annual International ACM SIGIR Conference on Research and Development in Information Retrieval, Association for Computing Machinery*, pp. 114–121.

[39] Yu, Li, Lu, Liu, and Xuefeng, Li (2005), 'A hybrid collaborative filtering method for multiple-interests and multiple-content recommendation in Ecommerce', *Expert Systems with Applications*, Vol. 28, pp. 67–77.

[40] Wang, Yi-Fan, Chuang, Yu-Liang, Hsu, Mei-Hua, and Keh, Huan-Chao (2004), 'A personalized recommender system for the cosmetic business', *Expert Systems with Applications*, Vol. 26(3), pp 427–434.

[41] Herlocker, Jon, Konstan, Joseph, Terveen, Loren, Lui, John C.S., and Riedl, T. (2004), 'Evaluating collaborative filtering recommender systems', *ACM Transactions on Information Systems*, Vol. 22, pp. 5–53.

[42] Montaner, M. et al. (2004), 'Evaluation of recommender systems through simulated users', *ICEIS*, Vol. 2004, pp. 303–308.

[43] Jing, Li-Ping, Huang, Hou-Kuan and Shi, Hong-Bo (2002), 'Improved feature selection approach TFIDF in text mining', *Proceedings of the International Conference on Machine Learning and Cybernetics*, Vol. 2, pp. 944–946.

[44] O. R. Zaiane (2002), 'Building a recommender agent for e-learning systems', *International Conference on Computers in Education*, Vol. 1, pp. 55–59.

[45] Mobasher, B., Dai, H., Luo, T. et al. (2002). Discovery and Evaluation of Aggregate Usage Profiles for Web Personalization. *Data Mining and Knowledge Discovery*, Vol. 6, pp. 61–82.

[46] Paik, Woojin, Yilmazel, Sibel, Brown, Eric, Poulin, Maryjane, Dubon, Stephane, and Amice, Christophe (2001), '*Applying natural language processing (NLP) based metadata extraction to automatically acquire user preferences*', In *Proceedings of the 1st international conference on Knowledge capture Association for Computing Machinery*, pp. 116–122.

[47] Sarwar, Badrul, Karypis, George, Konstan, Joseph, and Riedl, John (2000), '*Analysis of recommendation algorithms for e-commerce*', In *Proceedings of the 2nd ACM conference on Electronic commerce Association for Computing Machinery*, pp. 158–167.

[48] Schafer, J.B., Konstan, J., and Riedl, J. (2001), 'Electronic commerce recommender applications', *Journal of Data Mining and Knowledge Discovery*, Vol. 5, pp. 115–152.

[49] Shin, C. K. and Doermann, D. (1999), 'Classification of document page images based on visual similarity of layout structures', *Electronic Imaging*, pp. 182–190.

[50] Breese, John S., Heckerman, David, and Kadie, Carl, (1998), *'Empirical analysis of predictive algorithms for collaborative filtering'*, In *Proceedings of the Fourteenth Conference on Uncertainty in artificial intelligence Morgan Kaufmann Publishers Inc.*, San Francisco, CA, pp. 43–52.

[51] Resnick, Paul, Iacovou, Neophytos, Suchak, Mitesh, Bergstrom, Peter, and Riedl, John (1994), *'Group Lens: an open architecture for collaborative filtering of netnews'*, In *Proceedings of the 1994 ACM conference on Computer supported cooperative work Association for Computing Machinery*, pp. 175–186.

[52] Goldberg, David, Nichols, David, Oki, Brian M., and Terry, Douglas (1992), 'Using collaborative filtering to weave an information tapestry', *Communications of the ACM*, Vol. 35(12), pp. 61–70.

[53] Small, H. (1973), 'Co-citation in the scientific literature: a new measure of the relationship between two documents', *Journal of the American Society for Information Science*, Vol. 24, pp. 265–269.

[54] https://www.analyticsvidhya.com/blog/2015/08/beginners-guide-learn-content-based-recommender-systems/ (accessed August 13, 2020).

8

COLLABORATIVE FILTERING TECHNIQUES

Algorithms and Advances

PALLAVI MISHRA AND SACHI NANDAN MOHANTY

ICFAI Foundation For Higher Education, Hyderabad,
Telangana, India

Contents

8.1 Overview of Collaborative Filtering

The availability of abundant information on the internet makes it difficult to assess the actual need of people in the domain of e-commerce, which compels the implementation of a personalised information selection model in online multimedia services, online shopping and social networking. Among the various information selection technologies, the recommendation system is one that has been significantly developed due to its wide application in e-commerce, provides personalised recommendations of the product that suits the taste of the users. The world-renowned companies who have proficiency in the IT domain, such as Google, Amazon, Flipkart and online movie recommendation systems such as Netflix use such recommender systems as a decision-maker.

Various filtering techniques have been devised to develop effective recommendation systems. Content Filtering and Collaborative Filtering are two of the major techniques which are known for their fundamental building approaches.

Content-based (CB) filtering techniques map the feature attributes of new items (target items) to that of those items that were already purchased or preferred highly by the customer previously. Similar attributes of target items are recommended to the user in order to draw their attention and make them biased towards the recommended items to go for purchasing.

The other method is Collaborative Filtering (CF), otherwise known as information filtering system, which basically makes automatic predictions (filtering) about the interests of a user by collecting preferences or taste information from many users (collaborating). The basis for collaborative filtering is that users who are clustered into one group with respect to their preferences over the items are most probably inclined to give very similar preferences or ratings for the new and future target items. Thus, this particular approach proves to be an effective method of recommendation system that makes user-specific recommendation on the basis of patterns of the ratings for the items without possessing any exogenous information of either the item or the user. Although there are plenty of well-established methods in operation in the e-commerce domain, attempts are still being made to strengthen the recommendation system in order to make the system more efficient.

The Collaborative Filtering system came into existence with its real perspective after the Netflix Prize competition held in the month of October 2006. An opportunity was created for the research community for the first time to assess the ratings of 100 million movies that attracted thousands of active participants, including researchers, developers, scholars and enthusiasts to work in this field, which rendered widespread and progress of the Collaborative Filtering system. The innovators started to study the various techniques of recommendation systems of each generation in order to bring about improvements in prediction accuracy.

Typically, the functionality of an information filtering system involves the user who conveys their interests by rating items (either explicitly or implicitly). The collaborative filtering system matches the target user's ratings (i.e. active user) against those of other users who have rated over the same items and thus tends to find the people with most "similar" tastes (i.e. neighbors). Once achieved, clustering is done based on similarity of preferences over the particular items. The information filtering systems then began to recommend those items that the similar users (i.e. neighbors of target user) have rated highly but not yet being rated by the active or target user.

In collaborative filtering techniques, the additional items preferred by an active user are predicted by referring to either explicit or implicit ratings over the items provided by the similar taste users. There are two possible outcomes in collaborative techniques. In the first case the biasness of user's liking towards item (i.e. preference) is evaluated in numerical value. Secondly, the recommendation list of the top N items which is mostly preferred by the users is projected [1] (Figure 8.1).

Though Collaborative Filtering strives to capture the user-item interaction, but the fact is that observed ratings are mostly due to the effects associated with either users (i.e., user bias) or items (i.e. item bias). The baseline prediction model captures those inherent biases associated with both user and items to establish non-personalized predictions. For example, a liberal user will have a tendency to rate almost every item above average that he/she has purchased, while a critical one will rate everything relatively low [2, 3].

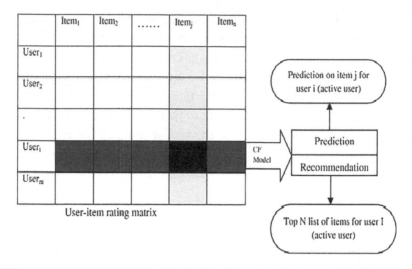

Figure 8.1   The collaborative filtering process

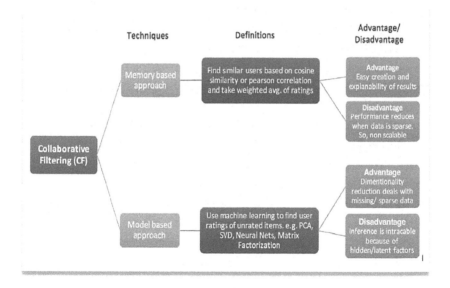

Figure 8.2   Types of collaborative filtering techniques

8.1.1 Approaches of Collaborative Filtering

For the sake of the initiation of the recommendation process, the Information Filtering Systems have to inherently compare the following two fundamental entities, i.e. users against items. With a view to facilitate the comparison, there exists two primary approaches which form the prime disciplines of Collaborative Filtering Systems (Figure 8.2).

- Memory-based Approach
- Model-based Approach

The best-known of all the Memory-based approaches is Neighborhood-based filtering. The Neighborhood-based algorithm calculates similarities between two users or items and makes a prediction for the user by taking the weighted average of all the ratings. The memory-based collaborative filtering approaches is broadly classified into two prime sections:

- User-User filtering
- Item-Item filtering

In the user-user-based collaborative filtering approach, the homogeneity in terms of preferences over items among the users is evaluated by comparing their ratings and taste patterns. After that, it determines the predicted rating for target items by the active user and accordingly recommend the most likely, i.e. the highly preferable target items those are already consumed and appreciated by the neighborhood users of the active user. In the case of item-based filtering techniques, the prediction is evaluated using similarities between items' features. In the case of each and every item those are rated by the active user, are retrieved and the correlation in terms of the features or attributes between the retrieved items and the target items are determined. The top N most correlated items are selected to predict the preference of the active user for the target item.

Recommendation systems based on the "Similarity-based" approach, i.e. either user-user-based similarity or item-item-based similarity filtering techniques are very intrinsic in nature and fairly simple to implement. This enriches the user experience due to its interpretability nature and motivates users by stimulating and finding a direct way to get connected with the system explicitly. In the case of the item-item-based filtering approach, there exists some scope to get updated recommendations of items, once new ratings are entered in the database of the personalized information model. However, its performance tends to decline while dealing with sparse matrix i.e. user-item rating matrix, having a maximum number of missing entries, which results in limiting the flexible identification of this approach and creates a complexity issue with large datasets.

To solve the sparsity matrix issue associated with the memory-based filtering technique, different types of Model-based filtering techniques, such as Bayesian Networks (Probabilistic model), Rule-based models such as the Association rule method, Classification models such as Clustering techniques and Artificial Neural Networks, and Dimensionality Reduction models such as Latent Factor Models have been proposed. All of these model-based techniques are popularly known for using a combination of data science techniques (data mining) and so-called 'soft' computing techniques (machine learning) in order to anticipate the taste pattern of an active user for an item. In addition, it can effectively handle the sparsity of data and reduce the cost computation using various dimensionality reduction techniques, such as Matrix Factorization, Singular Value Decomposition (SVD), Probabilistic Matrix Factorization, Bayesian Probabilistic Matrix Factorization, Low Rank Factorization, Non-negative Matrix Factorization, and Latent Dirichlet Allocation. Nevertheless, this model-based filtering technique significantly gives rise to a more precise dataset (i.e. a dataset with latent factors) to deal with having dimensionality reduction features, which can be treated comparatively easier rather than dealing with the whole dataset in case of memory-based technique.

Mostly, the Latent Factor Models are based on the Singular Value Decomposition (SVD) concept, which is popularly known for its low-rank optimization of the matrix by reducing the dimensionality factors. The basic working principle behind the latent factor model is that it maps user and item entities onto a latent feature space, based on their preferences over those items. In other words, those users that are heavily inclined to a particular item will be positioned near to that item in the latent feature space, and vice versa. The original user-item dataset matrix is decomposed into two matrices (the user feature matrix and the item feature matrix) as the dot product of the two vectors, i.e. the user feature vector and the item feature vector, will have a corresponding approximated value of the original rating value (explicitly provided by the users) of the dataset using the squared error minimization technique of the machine learning model. Thus, it will be quite useful for estimating the unknown rating values of a target item by the active user.

The asset of implementing an SVD-based Model is that it allows the integration of other sources of user information implicitly through incorporating the extended version of SVD, i.e. the SVD++ model, which enhance the accuracy of the prediction. Finally, we deal with a Time-Aware Factor Model that addresses temporal dynamics and transient effects of both users and items for better tracking user behaviour [1, 2].

8.1.2 Challenges

The common challenge of collaborative filtering is to deal with huge volumes of data in order to make precise and correct recommendations. The key challenges of collaborative filtering are:

1. **Data sparsity:** Most of the entries in the user-item matrix are sparsely rated (i.e. missing values), which leads to poor predictions because of the inadequate availability of information.
2. **Scalability:** The Traditional Collaborative filtering algorithms confront serious scalability problems due to the rampant growth of users and items in online networking; thus, the performance tends to decline with increasing scaling data due to computational constraints and complexity.
3. **Synonyms:** The same or similar items are sometimes named differently and it is not possible on the part of a recommender system to discover such latent association between the items and thus liable to recognize those items differently.
4. **Gray sheep:** Most of the times, the opinions of some users do not consistently agree or disagree with any group of people (i.e. peculiar or drift nature) and thus do not get benefits from collaborative filtering recommendation systems.
5. **Shilling attacks:** Since the recommendation system is accessible to everyone for rating, people may possess a biasness or may be inclined more towards their own products by rating them highly without any true and real facts and supposed to throw negative ratings of the items owned by their competitors or rivals in market. This stands as a barrier for the collaborative filtering system in making accurate predictions.
6. **Diversity and the long tail:** In the case of limited availability of users' past history, CF algorithms tend to recommend the products bias to popularity, which may result in poor personalised user specific recommendations.

8.1.3 Advances in Collaborative Systems

The most common approach of collaborative filtering technique is memory-based. The Neighbourhood approach is the best-known Memory-based Filtering techniques, which uses similarity measures for finding similar user or items to predict unknown ratings. Since no optimization technique is implemented in the neighborhood approach, it is very simple to use. However, its performance tends to decline while handling sparse data. Another disadvantage of the neighborhood approach is the computational costs in finding the appropriate similar items and similar users. Thus, this results in poor time complexity as well as space complexity. Moreover, as alternative to the memory based approach, in model-based collaborative filtering techniques, several dimensional reduction techniques, such as **SVD** (Singular Value Decomposition) are used to handle such sparsity and hence it outperforms the memory-based filtering techniques. The model-based filtering techniques can be further enhanced by incorporating other sources of user feedback, such as implicit ratings by using a more derived form of Matrix Factorization model, i.e. **SVD++**, which enables the further improvement of prediction accuracy. However, in order to improve the prediction accuracy various aspects of data are taken into consideration such as Temporal Dynamics Effects. In other words, user preferences over the items tend to change over the time that influence the user's behavior. Therefore, it is necessary to capture all of those temporal effects (time-drifting nature) and transient effects (short-lived) to make the model more accurate and relevant using time **SVD++**. On comparing these three models-based approaches, the accuracy level gets enhanced on adding additional sources of information about users and items, thereby outperforming the memory-based system in terms of handling the sparsity and giving more accurate results [1].

The memory-based model can be further modified by incorporating the global optimized methods which bring great improvements by facilitating two new possibilities, such as the development of a factorized neighborhood model which on implementation, leads to a great improvement in computational efficiency. The second possibility is the addition of Temporal Dynamics into the memory-based model which leads to better prediction accuracy. Thus, incorporating such Global Optimization Techniques results in a significant improvement in predicting accuracy in memory-based models, much as in the Matrix factorization-based model. The advances in both model-based and memory-based models will be discussed in later sections which describe further enhancements by adding various aspects of data.

8.2 Preliminaries

The two fundamental entities involved in a CF system are user and item. We have used special indexing letters such as users (u, v) and items (i, j) to represent them without any ambiguity while working with symbolical representations. A Collaborative Filtering system is incorporated with a rating matrix R, which consists of a list of m-users and n-items (let's say), where the entries are filled up by the ratings (i.e. preferences). A rating $r_{u,i}$ denotes the rating given by user u for item i, where higher value indicates strong preference and vice versa. We use the notation $\widetilde{r_{u,i}}$ for predicted rating value to distinguish from actual rating i.e. $r_{u,i}$.

In fact, one can even opt for different time units (such as in hours, days, etc) for rating which needs to be taken into account for more accurate prediction. The scalar $t_{u,i}$ denotes the time of rating $r_{u,i}$. Practically, most of the rating data entries are missing or sparse, due to the fact that the user in general tends to rate only a small portion of the data. for e.g., in Netflix data, almost 99% of the all possible ratings are absent.

The (u, i) pair for which ratings are known are stored in the set K = {(u,i)/$r_{u,i}$ is known}. Since, in a collaborative filtering-based recommendation system, we deal with user-item interaction, as such each user u is correlated with a set of items denoted by R(u), which holds all those items which are already rated by the user u. Similarly, R(i) signifies the set of customers or users who have rated item i. Since explicit feedback (i.e. preference shown on rating) is not always possible, in that case we use a notation N(u), which contains all implicit user feedback (i.e., browsing history, favourites, clicked items, etc.) [2, 3].

However, the aim is to fill up the missing entries while dealing with the sparse rating matrix. Therefore, both the nearest neighbors and the matrix factorization methods are considered in order to evaluate their respective predictive performance. In general, most collaborative filtering methods start with a simple baseline model, initiate the advancements in collaborative filtering by implementing Baseline Estimation Model for simpler understanding.

However, statistical or mathematical accuracy metrics will be used as the benchmark for the evaluation of the performance of filtering techniques in terms of prediction accuracy by differentiating the predicted rating and the original user rating. Of these statistical evaluation metrics, Root Mean Square Error (RMSE) is widely used for the purposes of evaluation. The formula for RMSE is given by:

$$RMSE = \sqrt{\frac{1}{n}\sum_{u,i}(r_{ui} - p_{ui})^2} \text{ , where } p_{ui} \text{ denotes predicted rating by}$$

user u on item I, r_{ui} denotes actual rating given by user u on item i and n is the total number of ratings in the user-item matrix dataset. The lower the value of RMSE, the greater the accuracy.

In order to build a Collaborative Filtering-based Recommendation System, models are trained initially to get fit well into the previously observed ratings. However, goal is to make the models more generalized such that it could be capable to predict future unseen data or unknown ratings. Therefore, the regularization concept is implemented to the learning parameters, whose magnitudes are penalized to control the weights or parameter values in order to avoid overfitting. In other words, regularization acts as a trade-off between the performance of recommendation system and the simplicity of the model. Regularization is controlled by penalizing the magnitudes of the learning parameters by using regularization parameter denoted by λ.

8.3 Baseline Prediction Model

In order to enhance the recommendation performance, Normalization is the predominant factor, which acts as an integral element for the predictor models. The baseline prediction method is useful for establishing non-personalized predictions against which personalized algorithms can also be compared for normalizing the data for use with more sophisticated algorithms.

Baseline estimation models capture the inherent biases incorporated with both users and items (i.e. item popularity), discounting the portion of user-item interaction signals, which contribute to the subjective evaluation (rating) of an item by the user. The baseline prediction model combines the overall average rating (μ), user bias (b_u), and item bias (b_i) for normalization [1, 3, 4].

A baseline prediction for an unknown rating $r_{u,i}$ is denoted by b_{ui} and accounts for the user and item bias effects.

Mathematically, the above concept can be formulated as follows:

$$b_{ui} = \mu + b_u + b_i \qquad (8.1)$$

where, μ = global average rating of the system

b_u = observed deviation of user u rating from the average

b_i = observed deviations of rating of item i by user u from the average

Suppose we want a baseline prediction for the rating of the movie "*Titanic*" by user "John." Now, let us say we have the overall average rating over all movies in the dataset, i.e. μ = 3.7 stars. Furthermore, the movie *Titanic* is marked higher than the average rating, so it tends to be rated (.i.e. b_i) 0.5 stars above the average rating. On the other hand, John is a critical user, who tends to rate 0.3 stars lower than the average rating.

Thus, the baseline prediction for *Titanic*'s rating by John would be 3.9 stars by calculating as follows:

$b_{ui} = \mu + b_u + b_i$, substituting the above value into this equation, we get:

$$b_{ui} = 3.7 \, 0.3 + 0.5 = 3.9$$

Let, $e_{u,i}$ represent error between actual rating and baseline rating prediction. In order to estimate b_u and b_i, solving the Least Squared Error problem of the objective function as follows:

$$min_{b*} \sum_{(u,i) \in K} \left(r_{u,i} - \mu - b_u - b_i \right)^2 + \lambda_1 \left(\sum_u b_u^2 + \sum_i b_i^2 \right) \qquad (8.2)$$

The regularization parameter λ_1 is used to avoid overfitting by penalizing the magnitudes of the learning parameters, such as b_u and bi. This least square problem will be solved efficiently by using Stochastic Gradient Descent (SGD), by taking derivative of the objective function and updating the weights iteratively till we reach the global optimum.

$$b_u \leftarrow bu - \gamma \cdot \frac{1}{2K} \frac{d}{d(bu)} \left(\sum_{(u,i)\in K} \left(r_{u,i} - \mu - b_i - b_u \right)^2 + \lambda_1 \left(b_i^2 + b_u^2 \right) \right)$$

$$\text{or}, b_u \leftarrow bu - \gamma \cdot \frac{1}{2K} \left(-2\left(r_{u,i} - \mu - b_i - b_u \right) + 2\lambda_1 \cdot bu \right)$$

$$\text{or}, b_u \leftarrow bu + \gamma \cdot \left(\left(r_{u,i} - \mu - b_i - b_u \right) - \lambda_1 \cdot bu \right)$$

$$\text{or}, b_u \leftarrow bu + \gamma \cdot \left(e_{u,i} - \lambda_1 \cdot bu \right) \tag{8.3}$$

Similarly,

$$b_i \leftarrow bi + \gamma \cdot \left(e_{u,i} - \lambda_1 \cdot bi \right) \tag{8.4}$$

The regularized parameters values such as λ_1 as can be obtained using cross-validation.

By decoupling the calculation of b_i from the calculation of b_u,

For each item i, we set,

$$b_i = \frac{1}{R(i)} \sum_{u\in R(i)} \left(r_{u,i} - \mu \right) \tag{8.5}$$

and for each user u,

$$b_u = \frac{1}{R(u)} \sum_{i\in R(u)} \left(r_{u,i} - \mu - b_i \right) \tag{8.6}$$

The baseline prediction techniques can be further improved and extended by considering temporal dynamics or time-based events within the data, which will be discussed in subsequent sections.

8.4 Model-Based Filtering with Advances

Model-based approach CF systems, such as Latent Factor Models, are the most widely adopted parameterized modelling techniques which decompose the rating matrix comprising of user and item interaction into two feature matrices, i.e. user feature and item feature matrix by using dimensionality reduction techniques. It results into an optimized low-rank matrix (low-dimensional matrix) form by transforming both user and item feature vector into the same latent feature space using several machine learning algorithms to build an optimized and accurate recommendation system. Some of the examples of latent factor models include: Probabilistic Latent Semantic Analysis (pLSA), Neural networks (NN), Latent Dirichlet Allocation (LDA), Matrix Factorization (also known as the SVD-based model). Recently, Matrix factorization techniques (MF) have become extremely popular in the recommender systems owing to its performance in terms of accuracy and high scalability factor (Figure 8.3).

In an information retrieval system, SVD (Singular Value Decomposition) plays a fundamental role in identifying latent semantic factors that affect the user's preferences. For example, in the case of movie recommendations, the latent factors (k) may include genre, directors, language, actor, etc. However, the implementation of SVD in the case of explicit user feedback (i.e. rating matrix) raises difficulties due to the sparsity of the dataset. Since most of the entries in the rating matrix have missing values (i.e. they are sparse), in this case SVD remains undefined. In other words, SVD is unable to account for the missing values or entries of any matrix. Therefore, instead of

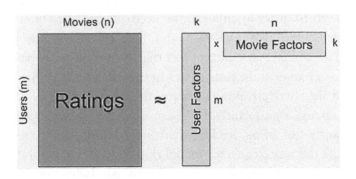

Figure 8.3   Matrix factorization technique

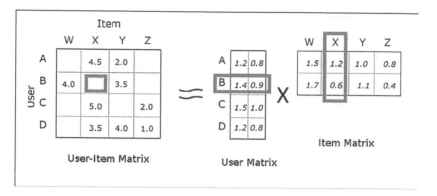

Figure 8.4 Decomposition of original matrix into two feature matrices using latent factor models

carelessly filling the missing values (i.e. imputation) which can be very expensive and may cause data distortion due to inaccurate imputation, modelling directly on the observed or known rating value would be quite significant while caution about evading from overfitting of the model by using a regularized model. With the help of machine learning techniques, our model can be built to predict the unknown ratings by evaluating the performance metrics by minimizing the Squared Error between the actual observed rating and the predicted rating. (Figure 8.4)

In this section, we will give details about various matrix factorization techniques, incorporating improved complexity as well as accuracy. First of all, we start with the elementary-based model "SVD" which forms a comprehensive understanding for matrix factorization techniques. Then, we exhibit the methodology in order to integrate additional source of user feedback (i.e. implicit) into the recommendation system in order to enhance the accuracy of prediction, through the implementation of "SVD++ model" (i.e. derivative model of SVD). Finally, we come across the real fact that customers' taste patterns for items may change as the time passes in the real world. With the emergence of new arrivals, product perception and popularity constantly tend to change significantly. Likewise, customer inclinations are also significantly unfolding, leading them to redefine their taste pattern. This leads the way to a factor model that addresses temporal dynamics effects to better capture the insights of user behaviour for further enrichment of the recommendation system [5, 6].

8.4.1 *Singular Value Decomposition (SVD)*

Matrix Factorization models often consider the task of recommendation as an optimization problem by using different dimensionality reduction techniques (such as SVD or PCA). It converts the original rating matrix R_{n*m} containing 'n users' for 'm items' into low-dimensional sparse matrix by factorizing it into two latent factor matrices, such as user factor matrix i.e. P_{n*k} and item factor matrix i.e. Q_{k*m}, where k<<m. Each element $r_{u,i}$ represents the rating of user u to the item i. The factorization is done in such a way that the rating matrix R is approximated as the inner product of two matrices such as P and Q (i.e. $\widetilde{R_{n*m}}$ = P_{n*k} * Q_{k*m}) and each observed rating $\widetilde{r_{u,i}}$ = qi^T. p_u (also called as predicted value). However, the term qi^T. p_u reflects the interconnection between the user *u* and the item *i*. In other words, it determines the preference of the users by looking at the features or attributes of the items.

The latent factor models represent both users and items as feature vectors characterizing their association with each of the latent factors. For example, in the case of movie recommendations, movies can be described in terms of features such as genre (comedy, drama, thriller, romantic), Thus, an item can be profiled as a vector enumerating the extent of possession of various features i.e. q_i, and an user can be embedded as a vector describing his/her affinity or inclination towards those as latent and a user can be modelled as a vector describing his/her affinity towards the corresponding item latent features such as p_u [6, 7, 8].

Ideally, the ratings are expected to be purely a function of the interaction between users' and items' latent factor vectors; however, in practice, they mostly get associated with effects of user bias and item bias. Thus, in addition to the pure interaction component, the actual ratings are also inflicted by certain biases which constitute the baseline component.

Thus, the formula for calculating the predicted rating is followed by the rule

$$\widetilde{r_{u,i}} = \mu + b_{i+}b_u + qi^T. p_u, \tag{8.7}$$

where, $\widetilde{r_{u,i}}$ = predicted rating value of the item i by user u.

μ = overall average rating

b_i = observed deviation of item i from the average

b_u = observed deviation of user u from the average

In order to learn the model parameters, we need to minimize the squared sum error (SSE) using the regularization model to avoid overfitting to gain ability to generalize new data in a more accurate manner.

Mathematically,

$$\min_{b*,q*,p*} \sum_{(u,i) \in K} \left(r_{u,i} - \mu - b_i - b_u - qi^T.pu \right)^2 + \lambda_2 \left(b_i^2 + b_u^2 + \| q_i \|^2 \| p_u \|^2 \right)$$

(8.8)

The constant λ_2 is known as a regularization parameter. This acts as a trade-off between predicted value and the actual value by penalizing the magnitudes of the learning parameters to avoid overfitting in case of complex models. The value of λ_2 is usually determined by using cross-validation. The model parameters can be calculated by using Stochastic Gradient Descent (SGD) to minimize the error function by simultaneously updating the learning parameters (i.e. weights) till we reach the global optima.

For each given actual rating $r_{u,i}$, corresponding predicted rating $(\widetilde{r_{u,i}})$ is estimated and the resultant prediction error is calculated as follows: error = actual value $(r_{u,i})$ – predicted value $\left(\widetilde{r_{u,i}}\right)$.

Let, say, e_{ui} represents the error between the actual and predicted value.

Mathematically,

$$e_{ui} = r_{u,i} - \widetilde{r_{u,i}}$$

(8.9)

For the given value of the actual rating in case of supervised learning $r_{u,i}$, we intend to modify the parameters or weights(b_i, b_u, p_u, q_i) in order to minimize the prediction error by using gradient descent method to converge towards the global minima by taking derivative of the error function as follows:

$$b_u \leftarrow bu - \gamma. \frac{1}{2K} \frac{d}{d(bu)} \left(\sum_{(u,i) \in K} \left(r_{u,i} - \mu - b_i - b_u - qi^T.pu \right)^2 + \lambda_2 \left(b_i^2 + b_u^2 + \| q_i \|^2 + \| p_u \|^2 \right) \right)$$

$$\text{or}, b_u \leftarrow bu - \gamma \cdot \frac{1}{2K}\left(-2\left(r_{u,i} - \mu - b_i - b_u - qi^T \cdot pu\right) + 2\lambda_2.bu\right),$$

$$\text{or}, b_u \leftarrow bu + \gamma \cdot \left(\left(r_{u,i} - \mu - b_i - b_u - qi^T \cdot pu\right) - \lambda_2.bu\right) \text{or},$$

$$b_u \leftarrow bu + \gamma \cdot \left(e_{u,i} - \lambda_2.bu\right)\left[\text{from eqs.}(8.7)\text{ and }(8.9)\right] \quad (8.10)$$

$$\text{Similarly}, b_i \leftarrow bi + \gamma \cdot \left(e_{u,i} - \lambda_2.bi\right) \quad (8.11)$$

$$q_i \leftarrow qi + \gamma \cdot \left(e_{u,i} - \lambda_2.qi\right) \quad (8.12)$$

$$p_u \leftarrow pu + \gamma \cdot \left(e_{u,i} - \lambda_2.pu\right) \quad (8.13)$$

Here, γ is known as the learning rate of the model. In this way, we can optimize the model to minimize the prediction error between actual rating and predicted rating value.

8.4.2 SVD++

The information domain of Collaborative Filtering consists of users those who have expressed their preferences over various items by giving either explicit feedback (i.e. rating) or implicit feedback. However, in practice, most of the entries in the user-item matrix of the datasets of the system remain sparse (i.e. there are missing values). In order to get relief from the sparsity problem and increase the accuracy of rating predictions, additional information about the users is to be incorporated using implicit user feedback information, such as browsing history, searching pattern, purchase history, mouse click movements, and so on.

Here, we focus on the SVD++ Model. In particular, the SVD++ Model adds the bias and implicit feedback of the user into the Matrix Factorization Model in order to improve the accuracy of the predicted rating. Since, SVD++ is a derived and conceptual model of SVD, it appends a factor term ($y_j \in R^k$) for each item, and these newly added item factors are meant to outline the features of the items those items are rated by users. Then the user's factor matrix, i.e. p_u, is modelled to capture users' affinity towards the item factors.

Thus, the rating prediction calculation in the SVD++ model is given as:

$$\widetilde{r_{u,i}} = \mu + b_{i+}b_u + qi^T\left(p_u + \left|N\left(u\right)\right|^{\frac{-1}{2}} \Sigma_{j\in N(u)}\ y_j \right) \quad (8.14)$$

where the first tier, $\mu + b_{i+}b_u$, describes the baseline properties of both the entities, i.e. item and user, without taking into consideration of any involved interactions between the entities. For example, let, say, it be stated that the "*Spiderman*" movie is known to be in the list of good movies, and the rating scale of the user "Tom", let us say, is above the average rating value of the movie. The next tier, $(qi^T\ p_u + \left|N\left(u\right)\right|^{\frac{-1}{2}} \Sigma_{j\in N(u)}\ y_j)$, yields the relationship or interlinkage between the user profile and the item profile and $\left|N\left(u\right)\right|^{\frac{-1}{2}} \Sigma_{j\in N(u)}\ y_j$ represents the perspective of implicit feedback. In the above-mentioned example, it may reflect that "*Spiderman*" and "Tom" are rated highly in the Animation scale.

To obtain the optimal model parameters (i.e. weights), the regularized square error function can be minimized. The objective function of SVD++ model is

$$min_{b*,q*,p*} \sum_{(u,i)\in K}\left(r_{u,i} - \mu - b_i - b_u - qi^T.\left(p_u + \left|N\left(u\right)\right|^{\frac{-1}{2}} \Sigma_{j\in N(u)}y_j \right) \right)^2$$
$$+ \lambda_3(b_i^2 + b_u^2 + \| q_i \|^2 + \| p_u \|^2) \quad (8.15)$$

Where λ_2 is the regularization parameter to avoid the overfitting by penalizing the magnitudes of the weights.

As earlier mentioned, $e_{u,i}$ represents the error between actual and the predicted values (i.e. $e_{ui} = r_{u,i} - \widetilde{r_{u,i}}$). *pu* is any element of the user matrix P, *qi* is any element of the item matrix *Q*.

Mathematically,

$$e_{ui} = r_{u,i} - (\mu + b_{i+}b_u + qi^T.\ (p_u + \left|N\left(u\right)\right|^{\frac{-1}{2}} \Sigma_{j\in N(u)}\ y_j)) \quad (8.16)$$

Model parameters can be learnt and modified by using the concept of the Gradient Descent Method which evaluates the predicted error e_{ui} for given actual rating $r_{u,i}$ and predicted rating $\widetilde{r_{u,i}}$. Subsequently, the user and item parameterized vectors along with baseline elements are updated and refined by moving in the direction towards global optima using stochastic gradient descent.

$$b_u \leftarrow bu - \gamma \cdot \frac{1}{2K} \frac{d}{d(bu)} \left[\sum_{(u,i)\in K} \left(r_{u,i} - \mu - b_i - b_u - qi^T \cdot \left(pu + \left| N(u) \right|^{\frac{-1}{2}} \Sigma_{j\in N(u)} y_j \right) \right)^2 \right. \\ \left. + \lambda_3 \left(b_i^2 + b_u^2 + \| q_i \|^2 + \| p_u \|^2 \right) \right],$$

$$\text{or, } b_u \leftarrow bu - \gamma \cdot \frac{1}{2K} \left[-2 \left(r_{u,i} - \mu - b_i - b_u - qi^T \cdot \left(pu + \left| N(u) \right|^{\frac{-1}{2}} \Sigma_{j\in N(u)} y_j \right) \right) \right. \\ \left. + 2\lambda_3 \cdot bu \right],$$

$$\text{or, } b_u \leftarrow bu + \gamma \cdot \left(\left[r_{u,i} - \mu - b_i - b_u - qi^T \cdot \left(pu + \left| N(u) \right|^{\frac{-1}{2}} \Sigma_{j\in N(u)} y_j \right) \right] - \lambda_3 \cdot bu \right),$$

$$\text{or, } b_u \leftarrow bu + \gamma \cdot \left(e_{u,i} - \lambda_3 \cdot bu \right) \tag{8.17}$$

Similarly,

$$b_i \leftarrow bi + \gamma \cdot \left(e_{u,i} - \lambda_3 \cdot bi \right) \tag{8.18}$$

$$q_i \leftarrow qi + \gamma \cdot \left(e_{u,i} \cdot (pu + \left| N(u) \right|^{\frac{-1}{2}} \Sigma_{j\in N(u)} y_j) - \lambda_3 \cdot qi \right) \tag{8.19}$$

$$p_u \leftarrow p_u + \gamma \cdot \left(e_{u,i} \cdot qi - \lambda_3 \cdot pu \right) \tag{8.20}$$

$$\forall j\in N(u): y_j \leftarrow y_j + \gamma \cdot (e_{u,i} \cdot \left| N(u) \right|^{\frac{-1}{2}} \cdot q_i - \lambda_3 \cdot y_j) \tag{8.21}$$

Since, in the real time scenario, explicit feedback from the users is not always available, therefore, introducing implicit feedback into the model deliberately by incorporating respective item factors will be a convenient approach which will impart more insightful knowledge into the recommendation system. For example, let's say user u exhibits a particular type of implicit preference to the items that is manifested in $N^1(u)$ (e.g., she browsed them), and may approach for some other kind of preferences over other items which is expressed in $N^2(u)$ (e.g., she purchased those items), and so on. We could implement this approach as follows:

$$\widetilde{r}_{u,i} = \mu + b_{i+}b_u + qi^T.\left(p_{u+}\left|N^1(u)\right|^{-1/2}\Sigma_{j\epsilon N(u)}\, y_j^{(1)} + \left|N^2(u)\right|^{-1/2}\Sigma_{j\epsilon N(u)}\, y_j^{(2)}\right)$$

(8.22)

The utmost importance of implicit feedback is relatively known in the absence of explicit user feedback in the user-item dataset. Upon adjusting the weights or parameters associated with the model, it will tend to learn the importance of other sources of feedback from the users explicitly and execute it in real time.

8.4.3 Time-Aware Factor Model

In reality, we often deal with dynamic data which often drift over the time. The analysis of such time-drifting data is useful to maintain a right balance between discounting less significant temporary effects on subsequent behaviour while capturing the long-term trends that reflects the intrinsic nature of data. This is otherwise known as "concept drift". In the field of Collaborative Filtering-based approach, the addition of temporal effects reflects the dynamic, time-drifting nature of user-item interaction, results in more accurate prediction due to the incorporation of other sources of information which adds time dimensions in rating data.

The modeling of temporal changes in user preferences plays a significant role in the design of the recommendation system. With the advent of new items or services, the focus of users gradually tends to change towards new arrivals. For example, users' preferences over the items vary depending upon fundamental factors, such as seasonal changes, weekends, festive seasons, and so on. However, many of the user behaviours changes due to the impact of circumstantial factors. For example, a change in the family structure or in the geographical position (migration) can exceptionally alter the shopping patterns which needs to be captured and reviewed for the updating of recommending items. Moreover, individuals gradually undergo alterations in terms of their preferences over recipes, attires, games, movies, music, etc. None of those changes can be captured by the methods that seek global concept drift. Instead, we need a more sensitive methodology which can capture time-drifting patterns in user behaviour and gain more meaningful insights to improve accuracy in the recommendation system [9].

For example, in a movie recommender system, users may change their preferred genre (i.e. comedy, drama, thriller, romantic, etc.) or may adopt a new perspective towards an actor or director. In addition, they may alter the appearance of their feedback. Let's say, in a system where users provide star ratings to products. At some point of time, a user that used to indicate a neutral preference by a "3 star" input, may now indicate dissatisfaction by the same "3star" feedback.

To improve our understanding about the importance of temporal models, let's take another instance. Suppose the members of a particular family are accessing the same account of any e-commerce site. Then the system cannot distinguish them separately, regardless of each member possessing their own inherent taste and preferences and the respective preference should get modelled accordingly. One way to get some distinction between different persons is by assuming that time-adjacent accesses are being carried out by the same member (sometimes on behalf of other members), which can be naturally captured by a temporal model that assumes the drifting nature of a customer.

The Netflix Challenge, held in October 2006, has practically proved the significance of temporal effects addition into the recommendation system. Two interesting temporal effects that emerged in the dataset released by Netflix are discussed below: One effect is an abrupt shift of rating scale (in early 2004), i.e., the mean rating value jumped from around 3.4 stars to above 3.6 stars. Another significant effect is that ratings given to movies tend to increase in line with the age of the movie. In other words, older movies receive higher ratings than newer ones.

Despite the fact that users do change their rating scale over the years, however, much of the old preferences are still helpful in analysing the historical and temporal patterns in order to isolate persistent signal from transient noise. This indeed helps to predict the future behaviour. Thus, just underweighting past actions tends to lose much relevant information. In other words, we seek a model that could explain user behaviour along the full extent of the time period, extracting signals from each time point.

One of the most successful approaches of CF is based on MF models, which lends itself to an adequate modelling of temporal effects. The decomposition of a rating into distinct portions is convenient here, as it allows us to treat different temporal aspects in separation. More specifically, we identify the following effects:

(1) User biases (b_u) change over time;
(2) Item biases (b_i) change over time;
(3) User preferences (p_u) change over time.

On the other hand, we would not expect a significant temporal variation of item characteristics (qi), as items, unlike humans, are static in their nature. We start with a detailed discussion of the temporal effects that are contained within the baseline predictors.

8.4.3.1 Time-Changing Baseline Parameters Temporal variation or deviation are incorporated in the baseline predictor models, through two major temporal effects. The first addresses the fact that an item's popularity may drift over time. For example, movies can swap in and out of popularity categories as triggered by external events such as appearance of an actor in a new movie, story based on patriotism, etc. This is manifested in this baseline model by treating them as item bias, i.e. b_i as a function of time. The second major temporal effect is based on the user bias that tends to change their preferences over items over time i. For instance, a user who used to rate an average movie "4 stars", may now rate such a movie "3 stars". Hence, in the baseline prediction model, we have taken into account user bias, i.e. b_u as a function of time. Inducing a time-sensitive baseline prediction for u's rating on item i at day t_{ui} in our model will be represented as:

$$b_{ui}(t) = \mu + b_u(t_{ui}) + b_i(t_{ui}) \qquad (8.23)$$

Here, function $b_{ui}(t)$ represents the baseline estimate for u's rating of i at day t and $b_u(t)$ and $b_i(t)$ are real valued functions that change over time.

A major consideration is needed to distinguish temporal effects that span over extended periods of time and more transient effects (i.e., short-lived effects) in order to check the rate of fluctuation on a timely basis. In case of movie rating, we don't expect movie likeability to fluctuate on a daily basis; it is more likely to change over more extend time periods. On the other side, we observe that user effects tend to change on a daily basis, reflecting instability and inconsistencies to natural behaviour, which requires a finer tune resolution to capture the transient effects.

We initiate our focus on time changing item bias $b_i(t)$, where item biases can split into time-based bins, using a constant item bias for each time period. For example, in a movie, the recommender system split the timeline into bins (i.e. small baskets) so as to balance the desire to achieve finer resolution (hence, smaller bins) with the need for enough ratings per bin (hence, larger bins):

$$b_i\left(t\right) = b_i + b_{i,\,\mathrm{Bin}(t)}. \tag{8.24}$$

However, it is more challenging in part of user side for binning the parameters. A higher resolution for users is expected to capture short-lived temporal effects with greater accuracy. Different function forms can be examined to parameterize temporal user behaviour or bias, with varying complexity and accuracy.

Thus, to capture a possible gradual drift of user bias, a simple modelling is built based on linear function. For each user u, we denote the mean date of rating by t_u. Now, if the user u rated a movie on day t, then the associated time deviation of this rating is defined as:

$$\mathrm{dev}_u\left(t\right) = sign|t - tu|.|t - tu|^{\beta}. \tag{8.25}$$

Here, $|t - tu|$ measures the number of days between dates t and tu. And the value of β can be set using the cross-validation method.

In order to formulate a time-dependent user bias, we introduce a new parameter for each user α_u i.e.,

$$b_u^{(1)}\left(t\right) = b_u + \alpha_u.\mathrm{dev}_u\left(t\right). \tag{8.26}$$

where α_u and b_u are two learning parameters.

A more flexible parameterization is offered by introducing splines. Let user u has given n_u ratings. We assigned k_u time points $\{t_1^u,...,t_k^u\}$, which has been spaced uniformly across the dates of u's ratings as kernels that control the following function:

$$b_u^{(2)}(t) = b_u + \frac{\sum_{l=1}^{ku} e^{-\sigma|t-tu|} b_{t_l}^u}{\sum_{l=1}^{ku} e^{-\sigma|t-tu|}} \tag{8.27}$$

The parameter $b_{t_l}^u$ is associated with the control points and are automatically learned from the data itself and the constant σ determines the smoothness of the spline.

In addition to gradual drift in the case of user bias, there may be a possibility of sudden change emerging as "spikes" associated with a single day or session. For example, in a movie rating dataset, it's been noticed that a user gives multiple ratings on a single day. Such an effect need not span more than a single day. This may reflect the instant mood of a user and its corresponding impact on the ratings given in that day, or changes in actual rater in the usage of multi-user accounts. To address such transient effects, a single parameter per user and day is assigned, i.e. $b_{u,t}$, absorbing the day-specific deviations which act as an additive component over the linear function-based simple model.

The time linear model thus becomes:

$$b_u^{(3)}(t) = b_u + \alpha_u.dev_u(t) + b_{u,t} \tag{8.28}$$

Similarly, to enhance the flexibility in parameterizing the model, the spline-based model is implemented as follows:

$$b_u^{(4)}(t) = b_u + \frac{\sum_{l=1}^{ku} e^{-\sigma|t-tu|} b_{t_l}^u}{\sum_{l=1}^{ku} e^{-\sigma|t-tu|}} + b_{u,t} \tag{8.29}$$

Thus, adopting various temporal dynamic effects of user-item bias in a single framework, the baseline predictor model is designed as such:

$$b_{ui} = \mu + b_u + \alpha_u.dev_u(t_{ui}) + b_{u,t_{u,i}} + b_i + b_{i,Bin(t_{ui})} \tag{8.30}$$

In order to learn the involved parameters (i.e. b_u, α_u, $b_{u,t}$, b_i, $b_{i,\,Bin(t)}$), we need to minimize the associated regularized square error using stochastic gradient descent (SGD) by updating the weights (or learning parameters) simultaneously.

$$min \sum_{(u,i)\partial K} \left(r_{u,i} - \mu - \alpha_u.dev_u\left(t_{ui}\right) - b_{u,t_{u,i}} - b_i - b_{i,Bin(t)} \right)^2$$
$$+ \lambda_4 (b_i^2 + b_u^2 + \alpha_u^2 + |\,b_{u,t_{u,i}}\,|^2 + |\,b_{i,Bin(t_{ui})}\,|^2) \qquad (8.31)$$

The first term in the above equation strives to get fit well into the training model, while the second term is incorporated with regularization parameter constant λ_5 to avoid overfitting by penalizing the magnitudes of the learning parameters.

So far, temporal dynamic effects have been discussed; this can be further extended by capturing other effects, i.e. incorporating periodic effects to improve the prediction accuracy. For example, some of the products or items may get popularity, depending upon the seasonal changes, festive occasions, etc. In the same way, periodic effects are also applicable on the user side. As such, users may possess different attitudes or buying patterns during the weekend as compared to working days. A way to model with those periodic effects is to dedicate a parameter for the combinations of time periods with item or users.

This way, the item bias becomes:

$$b_i\left(t\right) = b_i + b_{i,Bin(t)} + b_{i,period(t)} \qquad (8.32)$$

For instance, if we try to capture the change of item bias with the festive occasions, then period(t) ϵ {New Year, Holi, Diwali, Christmas}.

Similarly, recurring user effects may be modeled by modifying the time linear model of user bias to be:

$$b_u\left(t\right) = b_u + \alpha_u.dev_u\left(t\right) + b_{u,t} + b_{u,period(t)} \qquad (8.33)$$

Interestingly, this basic model incorporated with various temporal and periodic effects of both user and item bias has built a capability to explain various data variability. It basically discussed about various temporal effects that affects the baseline predictors without the involvement of the user-item interaction. At the same time, it's also necessary to focus on how does it affect the user likeability and their interaction with the items to be discussed next.

8.4.3.2 Time-Changing Factor Model   The way the time factor affects the baseline predictors, as a result of which the likelihood of change in preference or taste over items gradually takes place. For instance, a fan of the "thriller" genre of movie may become a fan of "comedy dramas" in the near future. Similarly, the user usually have a tendency to change their perception about actor or director. Such type of progression can be modeled by adding a user factor (p_u) as a function of time.

We can model each component of user preferences $p_u(t)^T = (p_{u1}(t),\dots$
$\dots,p_{uf}(t))$ in the same manner as we deal with user bias in the time linear model which leads to the following:

$$p_{uk}(t) = p_{uk} + \alpha_{uk}.\text{dev}_u(t) + p_{uk,t}. \tag{8.34}$$

Here, p_{uk} captures the static portion of the user bias factor
   $\alpha_{uk}.\text{dev}_u(t)$ approximates the portion of data that varies linearly with time.
   $p_{uk,t}$, absorbs the local drift concept i.e. day-specific changeability.

Finally, we can now encapsulate all the concepts that we have discussed to extend the SVD++ Model by integrating the time-changing parameters. This leads to a model called time SVD++, where the prediction rule can be formulated as follows:

$$\widetilde{r_{u,i}} = \mu + b_i\left(t_{u,i}\right)_+ b_u\left(t_{u,i}\right) + qi^T.\left(p_u\left(t_{u,i}\right)_+ \left|N\left(u\right)\right|^{\frac{-1}{2}} \Sigma_{j \in N(u)}\, y_j\right)$$

$$\tag{8.35}$$

Learning the model parameters can be done by minimizing the squared error function on the observed or known rating sets using regularized stochastic gradient descent. The procedure of determining the parameter values are analogous to the SVD++ model that is already discussed.

8.4.4 Comparison of Accuracy

In this section, we compare the accuracy of the three models, SVD, SVD++, and Time SVD++, that we have discussed in detail in the previous sections. All of the three methods are compared over a range of dimensionality factors (f) in terms of prediction accuracy measured in RMSE evaluation metric. SVD, being a basic model factorization model, provides intuition of the workflow of dimensionality reduction of features of the vast dataset. In addition, it also handles the sparsity of rating matrix by allowing the factorization of the original matrix into the item-based matrix and other one is the user-based matrix, thus calculating the prediction of unknown rating using dot product of the user-item interaction matrix. By contrast, SVD++, a derived factorization model, also leads to accuracy gain through the incorporation of an implicit user feedback outperforms SVD. Similarly, the time SVD++ model, which addresses the temporal effects of the data, also achieved enough accuracy to outperform the SVD++ Model. Thus, it seems, that the addition of various aspects of the data into the model results in an increase in the level of accuracy [1].

8.5 Memory-Based Model with Advances

The memory-based model is the most predominant approach of the Collaborative Filtering System. The memory-based Collaborative Filtering system can be implemented by either of the two main filtering techniques: user-based or item-based techniques.

In the user-based approach, similarities between the users (i.e. between active user and his neighbors) are calculated based on their rating behaviour and then, it computes the prediction of rating for a new item (that has been rated high by the neighborhood users) by the active user that hasn't been previously rated by him. Similarly, in an item-based filtering technique, all of the items already rated by the active user are retrieved to determine how similar the rated items are to the target item such that it could estimate the unknown rating of the target item by the same user.

As compared to the user-based CF technique, the item-based similarity approach is inherently more meaningful due to the fact that it possesses less temporal variation and more stability pertaining to their individual characteristics. Since the users remain more familiar with the items previously rated by them, as compared to their neighbors or likeminded users. Thus, it becomes on the part of the item-based approach more compliant in order to explain the reason behind the recommended, thereby enhancing the interpretability.

By contrast, in the user-based system, every individual possesses a wide variety of features which tends to change over time, resulting in instability and inconsistency which affects the similarity measures among them. For example, let's consider different genres of music such as classical, rock or pop. It may be possible that a user might like both acid rock and classical at the same time, but it's very rare that an item would belong to both the genres of music. Our focus will lie mostly on item-item approaches; however, we will implement advanced techniques to improve the prediction accuracy using the user-user approach.

Though most of the model-based approach (i.e. latent factor models) are widely preferred for handling sparsity of data and scalability, also express better aspects of data using various dimensional reduction techniques to predict unknown ratings. However, neighborhood models is also used by most of the top online recommenders such as Amazon, TiVo implements item-item-based collaborative filtering, which scales to massive data sets and produces high-quality recommendations in real time. The intuitions behind implementing the neighborhood model in real life include its simplicity, and its ability to explain the reason behind the recommended items which will make users more reliable towards the recommendations of items. Most importantly, they can handle new users after their first feedback on the system to recommend their personalized items.

Now, we will discuss about the measures to estimate the similarities between two items, which forms a basic fundamental block in neighborhood filtering techniques.

8.5.1 Calculation of Similarity Measures

Item-item-based similarity is calculated by taking the ratings of the users who have rated both the items. There are various mathematical formulations that can be implemented to calculate the similarity between two items (or users). Two of the most popular similarity measures are cosine-based and Pearson correlation coefficients.

Pearson correlation coefficients are used to measure the tendency of users to rate the items i, j in a similar manner. It is calculated as follows:

$$S(i,j) = \frac{\sum_{u \in U(i,j)} \left(r_{u,i} - b_{u,i}\right)\left(r_{u,j} - b_{u,j}\right)}{\sqrt{\sum_{u \in U(i,j)} \left(r_{u,i} - b_{u,i}\right)^2 \sum_{u \in U(i,j)} \left(r_{u,j} - b_{u,j}\right)^2}} \tag{8.36}$$

where $U(i,j)$ is the set of users who have rated both items i and j.

Here, we have subtracted the baseline prediction from the actual rating value to compensate user and item bias. Again, after similarity computation, we will now go for the prediction of unknown ratings.

Our goal is to predict the unknown ratings, i.e. $\widetilde{r_{u,i}}$, by user u on item i. Using a similarity measure, we found k-similar items rated by user u and this set of k-neighbors are denoted by $S^k(i;u)$.

The predicted rating can be calculated as a weighted average rating of neighboring items by adjusting the user and item effects through baseline predictors, i.e.

$$\widetilde{r_{u,i}} = b_{ui} + \frac{\sum_{j \in s^k(i;u)} S(i,j)\left(r_{u,j} - b_{u,j}\right)}{\sum_{j \in s^k(i;u)} S(i,j)} \tag{8.37}$$

In this above equation, there exists a dual role of similarity, i.e. identification of neighbors and act as an interpolation. However, it raises a concern regarding the introduction of interpolation weights. Let, say, an item has no significant neighbors rated by the user u; in that case, one of the simplest approaches is to ignore the neighborhood information, retaining only baseline predictors. Nevertheless, weighted average ratings of the uninformative neighbors are taken into account for predicted rating calculation.

Thus, this issue can be alleviated by using a regularization constant λ_5 in order to penalize the neighborhood portion in the case of insufficient neighborhood information. Thus, the predicted rating equation is further modified as follows:

$$\widetilde{r_{u,i}} = b_{ui} + \frac{\sum_{j \in s^k(i;u)} S(i,j)(r_{u,j} - b_{u,j})}{\lambda_5 + \sum_{j \in s^k(i;u)} S(i,j)}. \tag{8.38}$$

Further, we will enhance this model by computing interpolation weights $\{\theta_{ij}^u/j \in s^k(i; u)\}$, that enable the best prediction of the form:

$$\widetilde{r_{u,i}} = b_{ui} + \sum_{j \in s^k(i;u)} \theta_{ij}^u (r_{u,j} - b_{u,j}) \tag{8.39}$$

Here, the interpolation weights were derived by minimizing the objective function by least square error and update the weights simultaneously by using the stochastic gradient descent. The interpolation weights account for interdependencies among the neighbors. Finally, the success of the neighborhood model depends on the choice of interpolation weights, which are used to estimate unknown ratings from the neighboring observed ones.

Moreover, the memory-based approach can be further enhanced by incorporating the concept of both the classical neighborhood approach and matrix factorization models.

8.5.2 Enrichment of the Neighborhood Model

The neighborhood-based model can be enriched with the collaboration of both principles of classical neighborhood models and matrix factorization models in order to retain the practical advantages of those classical-based models with improved accuracy, as with matrix factorization models. Thus, incorporating global optimized techniques in the neighborhood approach is expected to enhance the accuracy level by raising new possibilities such as factorized model and temporal dynamics.

8.5.2.1 Global Neighborhood Model Mostly neighborhood methods are limited to small subsets of related ratings. This contrasts with matrix factorization, which strives to characterize the items and users using various dimensionality reduction techniques. In order to enhance the neighborhood model, we introduce global optimization techniques [1, 6].

An initial sketch of the model describes each rating r_{ui} by the equation:

$$\widetilde{r_{u,i}} = b_{ui} + \sum_{j \in R(u)} \left(r_{u,j} - b_{u,j}\right) w_{ij} \tag{8.40}$$

Earlier, we used to interpret weights (w_{ij}) as interpolation coefficients relating unknown ratings to the existing values in a neighborhood model. In order to facilitate global optimization, we no longer interpret weights to be interpolation coefficients, instead we take weights as a trade-off between the actual and predicted rating value, added to the baseline predictors. As such, the weight is the extent by which we increase the baseline prediction of r_{ui}, based on the observed value of $r_{u,j}$. For two related items, i and j, the value of w_{ij} will be expected as high value. Thus, whenever user u has rated item i higher than the expected or predicted rating value, in that case, $(r_{uj} - b_{uj})$ will be high. We would like to increase our estimate for u's rating for item i by adding $(r_{u,j} - b_{u,j})w_{ij}$ to the baseline prediction of r_{ui}.

In fact, the estimate rating will not deviate much from the baseline rating of item j by user u in case the user has rated as expected, which tends $(r_{u,j} - b_{u,j})$ around zero, or by an item j that is not known predictive to item I; thus, w_{ij} is close to zero.

We can use the binary user input, which was earlier found useful in MF models. Analysing the items that were rated regardless of rating value, by adding a weight parameter, and rewrite the above equation as follows:

$$\widetilde{r_{u,i}} = b_{ui} + \sum_{j \in R(u)} \left(r_{u,j} - b_{u,j}\right) w_{ij} + c_{ij} \tag{8.41}$$

Similarly, we can incorporate implicit user feedback, N(u) as such:

$$\widetilde{r_{u,i}} = b_{ui} + \sum_{j \in R(u)} \left[\left(r_{u,j} - b_{u,j} \right) w_{ij} + \sum_{j \in N(u)} c_{ij} \right] \qquad (8.42)$$

Here, w_{ij} and c_{ij} are the offsets added to the baseline predictors.

The implicit preference by user u on item j leads to an adjustment of the estimation $\widetilde{r_{u,i}}$, by c_{ij}. Thus, the value of c_{ij} will be high if item j is predictive on i.

Now, normalizing the parameters lead to:

$$\widetilde{r_{u,i}} = \mu + b_{i+}b_u + \left| R(u) \right|^{-\alpha} \sum_{j \in R(u)} \left[\left(r_{u,j} - b_{u,j} \right) w_{ij} + C_{ij} \right] \qquad (8.43)$$

where α controls the extent of normalization.

Let's denote $s^k(i)$ be a set of k most similar items to i. Additionally, we use $R^k(i; u) \stackrel{\text{def}}{=} R(u) \cap s^k(i)$.

$$\widetilde{r_{u,i}} = \mu + b_{i+}b_u + \left| R^k(i;u) \right|^{-1/2} \sum_{j \in R^k(i;u)} \left(r_{u,j} - b_{u,j} \right) w_{ij} + C_{ij} \qquad (8.44)$$

A major design goal of the new neighborhood model has been facilitated to bring an efficient global optimization procedure which prior neighborhood models were lacking. Thus, model parameters are learnt by solving the regularized least square problem as follows:

$$\min_{b*,q*,p*} \sum_{(u,i) \in K} \left(r_{u,i} - \mu - b_i - b_u - \left| R^k(i;u) \right|^{-1/2} \sum_{j \in R^k(i;u)} \left(\left(r_{u,j} - b_{u,j} \right) w_{ij} + C_{ij} \right) \right)^2$$

$$+ \lambda_5 \left(b_i^2 + b_u^2 + \sum_{j \in R^k(i;u)} w_{ij}^2 + C_{ij}^2 \right). \qquad (8.45)$$

The finest fitting solution of this convex problem can be solved efficiently using stochastic gradient descent. Let us denote $e_{ui} = r_{u,i} - \widetilde{r_{u,i}}$.

For a given training rating $r_{u,i}$, we modify the learning parameters $(b_i, b_u, w_{ij}, C_{ij})$ by moving in the opposite direction of the gradient until we reach the global optimum value, iterating over all known ratings in K.

$$b_u \leftarrow b_u - \gamma \cdot \frac{1}{2K} \frac{d}{d(b_u)} \left(\left[\sum_{(u,i)\in K} \left(r_{u,i} - \mu - b_i - b_u - \left| R^k(i;u) \right|^{-1/2} \sum_{j\in R^k(i;u)} \left(\left(r_{u,j} - b_{u,j} \right) w_{ij} + C_{ij} \right) \right)^2 \right. \right. $$
$$\left. \left. + \lambda_5 \left(b_i^2 + b_u^2 + \sum_{j\in R^k(i;u)} w_{ij}^2 + c_{ij}^2 \right) \right] \right),$$

$$\text{or, } b_u \leftarrow b_u - \gamma \cdot \frac{1}{2K} \left(-2 \left(r_{u,i} - \mu - b_i - b_u - \left| R^k(i;u) \right|^{-1/2} \sum_{j\in R^k(i;u)} \left(\left(r_{u,j} - b_{u,j} \right) w_{ij} + C_{ij} \right) \right) + 2\lambda_5 \cdot b_u \right),$$

$$\text{or, } b_u \leftarrow b_u + \gamma \cdot \left(\left(r_{u,i} - \mu - b_i - b_u - \left| R^k(i;u) \right|^{-\frac{1}{2}} \sum_{j\in R^k(i;u)} \left(\left(r_{u,j} - b_{u,j} \right) w_{ij} + C_{ij} \right) \right) - \lambda_5 \cdot b_u \right),$$

$$\text{or, } b_u \leftarrow b_u + \gamma \cdot \left(e_{u,i} - \lambda_5 \cdot b_u \right)$$

similarly,

$$b_i \leftarrow b_i + \gamma \cdot \left(e_{u,i} - \lambda_5 \cdot b_i \right) \qquad (8.46)$$

$$\forall j \in : R^k(i;u),$$

$$w_{ij} \leftarrow w_{ij} + \gamma \cdot \left(e_{u,i} \cdot \left| R^k(i;u) \right|^{-1/2} \sum_{j\in R^k(i;u)} \left(\left(r_{u,j} - b_{u,j} \right) - \lambda_5 \cdot w_{ij} \right) \right)$$
$$(8.47)$$

$$c_{ij} \leftarrow c_{ij} + \gamma \cdot \left(e_{u,i} \cdot \left| R^k(i;u) \right|^{-1/2} - \lambda_5 \cdot c_{ij} \right) \qquad (8.48)$$

The Meta parameters γ (step size) and λ_9 are determined by cross-validation. Another important parameter is k, which controls the neighborhood size. An increase in k shows a greater accuracy result in the test case. Hence, the choice of k will reflect a trade-off between prediction accuracy and computational cost.

In order to relax the constraint in the neighborhood size limit, we will further enhance the model by incorporating the factored version model. This would eliminate this trade-off by allowing us to work with the most accurate k = ∞ while minimizing running time.

For each user u and item iϵ R(u) in eq. (8.43), we need to modify both $\{w_{ij}, c_{ij}/j\epsilon$ R(u), i.e. overall time complexity of the training phase is O $(\sum_u |R(u)|^2)$

8.5.2.2 A Factorized Neighborhood Model

Since, in a global optimized-based neighborhood model, the time complexity and space complexity are, respectively, O $(\sum_u |R(u)|^2)$ *and* $O(m + n^2)$. Here, m denotes the number of users and n denotes the number of items. Here, using the factorized neighborhood concept, we could improve time and space complexity by sparsifying the model through pruning unlikely item-item relations. Sparsification is controlled by the parameter $k \leq n$, which reduced running time and allowed space complexity O(m+nk). The accuracy of the model prediction tends to decrease for a smaller number of k; thus, we will enhance the model by retaining the accuracy and, at the same time, allowing significant decline in time complexity and space complexity.

8.5.2.2.1 Factoring Item-Item Relationships Factorizing item-item relationships by incorporating each item i with three virtual vectors, i.e. q$_i$, x$_i$, y$_j$ $\in R^f$. We can assign qi^T x$_i$ to w_{ij}, likewise, we impose qi^T. y$_j$ to C$_{ij}$. These vectors attempt to map items to f-dimensional latent factor space where they are measured against several aspects that are disclosed automatically through learning phase of training data.

Now, substituting the values in place of w_{ij} and C$_{ij}$ in eq. (8.43), we get,

$$\widetilde{r_{u,i}} = \mu + b_{i+} b_u + \left|R(u)\right|^{-\frac{1}{2}} \sum_{j \in R(u)} \left[\left(r_{u,j} - b_{u,j}\right) qi^T . x_j + qi^T . y_j \right]$$

$$\text{Or,} \widetilde{r_{u,i}} = \mu + b_{i+} b_u + qi^T . \left(\left|R(u)\right|^{-\frac{1}{2}} \sum_{j \in R(u)} \left(r_{u,j} - b_{u,j}\right) x_j + y_j \right) \quad (8.49)$$

The term $\left(\left|R(u)\right|^{-\frac{1}{2}}\sum_{j\in R(u)}\left(r_{u,j}-b_{u,j}\right)x_j+y_j\right)$ depends only on the user u being independent on item.

Now, minimizing the objective function using regularized least square error as such:

$$\min_{b*,q*,p*}\sum_{(u,i)\in K}(r_{ui}-\mu-b_i-b_u-qi^T.(\left|R(u)\right|^{-\frac{1}{2}}\sum_{j\in R(u)}\left(r_{u,j}-b_{u,j}\right)x_j+y_j))^2$$

$$+\lambda_{10}(b_i^2+b_u^2+\|q_i\|^2+\sum_{j\in R(u)}\|x_j\|^2+\|y_j\|^2)$$

$$(8.50)$$

Optimization is done using the stochastic gradient descent as follows:

$$p_u\leftarrow(\left|R(u)\right|^{-\frac{1}{2}}\sum_{j\in R(u)}\left(r_{u,j}-b_{u,j}\right)x_j+y_j)\qquad(8.51)$$

$$b_u\leftarrow bu+\gamma.\left(e_{u,i}-\lambda_{10}.bu\right)\qquad(8.52)$$

$$b_i\leftarrow bi+\gamma.\left(e_{u,i}-\lambda_{10}.bi\right)\qquad(8.53)$$

$$q_i\leftarrow qi+\gamma.\left(e_{u,i}.pu-\lambda_{10}.qi\right)\qquad(8.54)$$

$$x_j\leftarrow x_j+\gamma.\left(e_{u,i}.qi^T.(\left|R(u)\right|^{-\frac{1}{2}}\sum_{j\in R(u)}\left(r_{u,j}-b_{u,j}\right)-\lambda_{10}.xj\right)\qquad(8.55)$$

$$y_j\leftarrow y_j+\gamma.\left(e_{u,i}.qi^T-\lambda_{10}.y_j\right)\qquad(8.56)$$

The time complexity of this factorized neighborhood model is linear with the size of input, i.e. $O\left(f.\sum_u |R(u)|\right)$, which is far better than the non-factorized neighborhood model that possesses time complexity $O\left(\sum_u |R(u)|^2\right)$. In the case of space complexity, the factorized model outperforms the non-factorized model. In the factorized neighborhood model, space complexity is reduced to O(m+nf), varies linearly to input size, whereas the non-factorized neighborhood model requires storing all pairwise relations between items, leading to a quadratic space complexity of O(m+n²).

Sparsification technique inevitably leads to a degradation in accuracy because of the occurrence of missing relations. Moreover, retaining only top-k similar neighbors in the storage in order to reduce space complexity, costs huge computational efforts.

However, all these issues do not exist in the factorized neighborhood model, which offer a linear time and space complexity without affecting accuracy.

8.5.2.2.2 A User-User Model The user-based neighborhood model predicts ratings by finding the neighborhood of the active user (i.e. like-minded users) who have rated highly the items those have not been previously rated by that active user. Such a model can be implemented as follows:

$$\widetilde{r_{u,i}} = \mu + b_{i+}b_u + |R(i)|^{-\frac{1}{2}} \sum_{v \in R(i)} (r_{v,i} - b_{v,i}) w_{uv} \tag{8.57}$$

The set $R(i)$ consists of all the users who have rated item i whereas user v belongs to the neighborhood of active user u.

As discussed earlier, item-item-based CF is most stable as compared to user-based as such users tend to change their interests much faster than the item. In the case of web articles, however, with news items which change drastically over time, here, user-user based CF recommendation system will fit well. Further, some of the issues faced in item-item filtering can be handled by using a user-based approach. For instance, suppose we want to predict the rating given by user on item I, i.e. r_{ui}. And let, say, the items previously rated by the user u be not so relevant to the item i in terms of similarity; in that case, the item-item-based neighborhood approach will fail to implement. But the user-user approach can work well in such types of real scenario. Thus, the user-user approach acts as a complement to the item-item-based filtering techniques [10].

However, the major limitation in user-based filtering technique lies in computational cost and effort. Since users tend to be higher in quantity than the number of items that exist in the system, pre-computing and storing each user-user pair is overly time-consuming and expensive, making it even completely impractical. In addition to the high space complexity, i.e. $O(m^2)$ the time complexity of the optimized model is O $(\sum_i |R(i)|^2)$, which is also higher than the item-item approach, since R(i) is expected to be much higher than R(u). Because of these issues, it has made user-user-based approach less practical.

The following two approaches can be implemented in order to overcome the above issues in terms of complexity.

1. Factorized User Model

The above computational issues can be avoided by factorizing the user-user model by associating each user u with two vectors, p_u, $z_u \in R^f$. Now we can assume the user-user relations can be modified as $w_{uv} = p_u^T . z_v$.

Let us substitute this value in place of w_{uv} in the above equation.

$$\widetilde{r_{u,i}} = \mu + b_{i+}b_u + \left|R(i)\right|^{-\frac{1}{2}} \sum_{v \in R(i)} \left(r_{v,i} - b_{v,i}\right) p_u^T . z_v.$$

$$\text{Or, } \widetilde{r_{u,i}} = \mu + b_{i+}b_u + p_u^T . \left|R(i)\right|^{-\frac{1}{2}} \sum_{v \in R(i)} \left(r_{v,i} - b_{v,i}\right) . z_v \qquad (8.58)$$

The above equation has transformed into an efficient computation by reducing the time complexity in linear time, i.e. O $\left(f . \sum_i \left|R(i)\right| \right)$ and space complexity O(n+mf), which linearly varies w.r.t input size.

Thus, factorizing the neighborhood model of the user-based approach has significantly reduced the time and space complexities. It therefore appears to be a well-implemented model that can match the speed and accuracy of an item-item model.

2. Fusing the Item-Item and User-User Model

Both item-item similarity and user-user similarity-based approaches have addressed several features or characteristics of the data individually. However, the overall level of accuracy is predicted to be intensified at fusing the factorized features of both the models.

Here, a model is devised that incorporates the features of both the item-item and the user-user models, leading to:

$$\widetilde{r_{u,i}} = \mu + b_{i+}b_u + |R(u)|^{-\frac{1}{2}} \sum_{j \in R(u)} \left[\left(r_{u,j} - b_{u,j} \right) qi^{T}. \, x_j + qi^{T}. y_j \right]$$
$$+ |R(i)|^{-\frac{1}{2}} \sum_{v \in R(i)} [\left(r_{v,i} - b_{v,i} \right). \, p_u^{\,T}. z_v] \tag{8.59}$$

Here the learning parameters are updated simultaneously using the stochastic gradient descent to obtain the global minima value in order to minimize the prediction error.

8.5.2.3 Temporal Dynamics at Neighbourhood Models Since, user preferences tend to drift over time, thus incorporating temporal effects into the CF models is of paramount importance in enhancing the accuracy of predictions. Now using the global optimized prediction rules, which has the advantage of integrating temporal effects, is implemented as follows:

$$\widetilde{r_{u,i}} = \mu + b_{i+}b_u + |R(u)|^{-\frac{1}{2}} \sum_{j \in R(u)} \left[\left(r_{u,j} - b_{u,j} \right) w_{ij} + C_{ij} \right] \tag{8.60}$$

Here, the first component, $= \mu + b_{i+}b_u$, denotes the baseline predictor portion used for handling the user and item bias effects. This component mostly explains the variability in the observed signal.

The second component, $\left| R(u) \right|^{-\frac{1}{2}} \sum\limits_{j \in R(u)} \left[\left(r_{u,j} - b_{u,j} \right) w_{ij} + C_{ij} \right]$, holds more informative signal which deals with user-item interaction.

In the baseline prediction part, adding the temporal effects replace b_i with $b_i(t_{ui})$ which represents item bias over temporal effects. Similarly, $b_u(t_{ui})$ which denotes the drifting nature of user bias.

In handling the temporal effects, the time-drifting nature of the data could have a possibility to suppress the longer association exists between items. So, those drifting data needs to be treated adequately. Therefore, initially we need to parameterize the decaying relations between two items rated by user u by adopting an exponential decay function, i.e. $e^{-\beta_{u.} \Delta t}$, where $\beta_u > 0$, controls the user-specific decay rate, the value of which will be learnt from the data itself, leads to the prediction rule as follows:

$$\widetilde{r_{u,i}} = \mu + b_i \left(t_{ui} \right) + b_u \left(t_{ui} \right) + \left| R(u) \right|^{-\frac{1}{2}} \sum_{j \in R(u)} e^{-\beta_{u.}\left(t_{ui} - t_{uj} \right)} \left[\left(r_{u,j} - b_{u,j} \right) w_{ij} + C_{ij} \right],$$

The involved learning model parameters, $b_i(t_{ui}) = b_i + b_{i, \text{Bin}(t_{ui})}$, $b_u(t_{ui}) = b_u + \alpha_u.\, \text{dev}_u(t_{ui}) + b_{u,\, tui}$, β_u, w_{ij}, C_{ij} are learnt by minimizing the associated regularized squared error as follows:

$$\min \sum_{(u,i) \in K} \left(\begin{array}{c} r_{u,i} - \mu - b_u - \alpha_u . dev_u \left(t_{ui} \right) - b_{u,t_{u,i}} - b_i - b_{i, \text{Bin}(t_{ui})} \\[2mm] - \left| R(u) \right|^{-\frac{1}{2}} \sum_{j \in R(u)} e^{-\beta_{u.}\left(t_{ui} - t_{uj} \right)} \left[\left(r_{u,j} - b_{u,j} \right) w_{ij} + C_{ij} \right] \end{array} \right)^2 \tag{8.61}$$
$$+ \lambda_{10} \left(b_i^2 + b_u^2 + \alpha_u^2 + \left| b_{u,t_{u,i}} \right|^2 + b_{i, \text{Bin}(t_{ui})}^2 + w_{ij}^2 + c_{ij}^2 \right)$$

The minimization of the objective function is done using the stochastic gradient descent. Here, the training phase time complexity is nearly the same as that of non-factorized neighborhood model (i.e. Global Neighborhood Model), i.e. O $(\sum_u |R(u)|^2)$. However, we could bring enhancements by improving time complexity and space complexity by bringing sparsity to the model through pruning using item-item relations.

Thus, incorporating temporal effects in the neighborhood model out-turns in a remarkable refinement in predictive accuracy, much like in the memory-based, i.e. matrix factorization, model.

8.6 Conclusion

Collaborative Filtering is the most predominant technique for recommendation systems. The two most familiar approaches of Collaborative Filtering include: memory-based filtering (i.e., similarity among the nearest neighbors) and model-based approach (i.e., matrix factorization model). Since the neighborhood approach deals with similarity measures in order to find the similar items or users of same interests, it costs a huge computational effort to find the neighborhood of items (users) from a vast dataset size; thus, it increases the time and space complexity. However, the model-based approach (i.e. SVD), being in its most elementary configured structure, signifies both items and users in the form of latent factors that describe the rating behaviour pattern using dimensionality reduction techniques. Further, model-based filtering techniques can be enhanced by incorporating indirect and unexpressed user feedback (i.e. implicitly), through the SVD++ Model which improves over SVD upon the addition of other sources of useful information about users. To date, the refinement or enhancement achieved by the time SVD++ Model over the SVD in terms of prediction accuracy of unknown ratings is typically more significant. The Neighborhood models also strive hard to overcome from its own constraints and limitations of holding high time and space complexity by incorporating the Global Optimized concept which leads to new possibilities such as the Factorized model and the treatment of Temporal dynamics, which leads to a great improvement in computational efficiency, thereby leading to better prediction accuracy.

References

1. Koren, Yehuda, and Robert Bell. "Advances in collaborative filtering." *Recommender systems handbook*. Springer, Boston, MA, 77–118 (2015).
2. Gogna, Anuprriya, and Angshul Majumdar. *Latent factor models for collaborative filtering*. Diss. IIIT-Delhi (2017).

3. Ma, Wenming, Junfeng Shi, and Ruidong Zhao. "Normalizing item-based collaborative filter using context-aware Scaled baseline predictor." *Mathematical Problems in Engineering* 2017 (2017).

4. Ekstrand, Michael D., John T. Riedl, and Joseph A. Konstan. *Collaborative filtering recommender systems*. Now Publishers Inc. (2011).

5. Isinkaye, F. O., Y. O. Folajimi, and B. A. Ojokoh. "Recommendation systems: Principles, methods and evaluation." *Egyptian Informatics Journal* 16(3): 261–273 (2015).

6. Koren, Yehuda. *"Factorization meets the neighborhood: A multifaceted collaborative filtering model."* Proceedings of the 14th ACM SIGKDD international conference on Knowledge discovery and data mining. (2008).

7. Zhou, Xun, et al. "SVD-based incremental approaches for recommender systems." *Journal of Computer and System Sciences* 81(4): 717–733 (2015).

8. Xian, Zhengzheng, et al. "New collaborative filtering algorithms based on SVD++ and differential privacy." *Mathematical Problems in Engineering* (2017).

9. Koren, Yehuda. *"Collaborative filtering with temporal dynamics."* Proceedings of the 15th ACM SIGKDD international conference on Knowledge discovery and data mining. (2009).

10. Shi, Wenchuan, Liejun Wang, and Jiwei Qin. "User Embedding for Rating Prediction in SVD++-Based Collaborative Filtering." *Symmetry* 12(1): 121 (2020).

Index

Page numbers in *italics* refer to figures and those in **bold** refer to tables.